SpringerBriefs in Business

AF842991

SpringerBriefs present concise summaries of cutting-edge research and practical applications across a wide spectrum of fields. Featuring compact volumes of 50 to 125 pages, the series covers a range of content from professional to academic. Typical topics might include:

- A timely report of state-of-the art analytical techniques
- A bridge between new research results, as published in journal articles, and a contextual literature review
- A snapshot of a hot or emerging topic
- An in-depth case study or clinical example
- A presentation of core concepts that students must understand in order to make independent contributions

SpringerBriefs in Business showcase emerging theory, empirical research, and practical application in management, finance, entrepreneurship, marketing, operations research, and related fields, from a global author community.

Briefs are characterized by fast, global electronic dissemination, standard publishing contracts, standardized manuscript preparation and formatting guidelines, and expedited production schedules.

Beatrice Re • Pietro Previtali
Editors

Tackling Malnutrition in a Complex World

Strategies, Economic and Organizational Impacts, and Implications

Springer

Editors
Beatrice Re
Department of Economics and Management
University of Pavia
Pavia, Italy

Pietro Previtali
Department of Economics and Management
University of Pavia
Pavia, Italy

ISSN 2191-5482 ISSN 2191-5490 (electronic)
SpringerBriefs in Business
ISBN 978-3-032-26437-4 ISBN 978-3-032-26438-1 (eBook)
https://doi.org/10.1007/978-3-032-26438-1

© The Editor(s) (if applicable) and The Author(s), under exclusive license to Springer Nature Switzerland AG 2026. This book is an open access publication.

Open Access This book is licensed under the terms of the Creative Commons Attribution-NonCommercial-NoDerivatives 4.0 International License (http://creativecommons.org/licenses/by-nc-nd/4.0/), which permits any noncommercial use, sharing, distribution and reproduction in any medium or format, as long as you give appropriate credit to the original author(s) and the source, provide a link to the Creative Commons license and indicate if you modified the licensed material. You do not have permission under this license to share adapted material derived from this book or parts of it.
The images or other third party material in this book are included in the book's Creative Commons license, unless indicated otherwise in a credit line to the material. If material is not included in the book's Creative Commons license and your intended use is not permitted by statutory regulation or exceeds the permitted use, you will need to obtain permission directly from the copyright holder.
This work is subject to copyright. All commercial rights are reserved by the author(s), whether the whole or part of the material is concerned, specifically the rights of translation, reprinting, reuse of illustrations, recitation, broadcasting, reproduction on microfilms or in any other physical way, and transmission or information storage and retrieval, electronic adaptation, computer software, or by similar or dissimilar methodology now known or hereafter developed. Regarding these commercial rights a non-exclusive license has been granted to the publisher.
The use of general descriptive names, registered names, trademarks, service marks, etc. in this publication does not imply, even in the absence of a specific statement, that such names are exempt from the relevant protective laws and regulations and therefore free for general use.
The publisher, the authors and the editors are safe to assume that the advice and information in this book are believed to be true and accurate at the date of publication. Neither the publisher nor the authors or the editors give a warranty, expressed or implied, with respect to the material contained herein or for any errors or omissions that may have been made. The publisher remains neutral with regard to jurisdictional claims in published maps and institutional affiliations.

This Springer imprint is published by the registered company Springer Nature Switzerland AG
The registered company address is: Gewerbestrasse 11, 6330 Cham, Switzerland

If disposing of this product, please recycle the paper.

Preface

Malnutrition is one of the key challenges that the contemporary world faces. While the term is traditionally linked to hunger and undernutrition, malnutrition actually includes three main forms, i.e., undernutrition, micronutrient deficiencies (namely, "hidden hunger"), and overnutrition/obesity/overweight. The latter is becoming an increasingly serious problem, as evidenced by the fact that the worldwide increase in malnutrition from 1990 to 2022 is mainly driven precisely by the rise of the obesity rate. The problem is even more serious in regard to children, as the percentage of school-age children and adolescents with obesity increased from 1.7% in 1990 to 6.9% in 2022. It is commonly thought that childhood overweight and obesity are issues that concern mostly industrialized countries, while they also concern developing countries, where the levels are increasingly becoming similar to those of some industrialized countries. As the economic, social, and psychological impacts of malnutrition are serious and long-lasting for individuals and their families, communities, and countries, addressing it by achieving food security, improving nutrition, and promoting sustainable dietary patterns is one of the most critical priorities of our century.

By acknowledging the multifaceted nature of malnutrition, this book is inspired by the idea that the topic needs to be addressed from different angles and through a multidisciplinary approach. We therefore adopt an innovative posture to the subject, which includes the analysis of multiple target groups through a threefold perspective: health and nutritional, economic, and organizational.

After an initial overview of the book's aims and structure, we introduce the first perspective regarding health and nutrition. We highlight our key element of novelty, which lies in considering not only all three main forms of malnutrition but also different target groups, which are all vulnerable targets (except for employees), i.e., individuals with low socioeconomic status (SES), children, autistic individuals, hospitalized patients, elderly people, and organization employees. Each target group is analyzed separately, as it presents specific features, needs, and nutrition-related risks that deserve a specific, thorough examination. By following this holistic approach, we address several research questions involving multiple disciplines. We focus our research on a developed country, i.e., Italy, where the spread and impact of disease-related malnutrition (DRM) in the Italian context are remarkable. Investigating the different aspects of malnutrition in this multifaceted research context could thus offer significant insights into the most vulnerable populations (including those that have been overlooked by the extant literature) and related practices to tackle the DRM.

The main macro-questions we address in this book are the following:

(a) What is the development process of malnutrition? From a health and nutritional perspective, which target groups are more vulnerable to being affected by malnutrition (defined as overweight or obesity, undernutrition, and/or micronutrient deficiencies), and how does it affect their health?
(b) What are the economic impacts of malnutrition, and which policies can be implemented to address malnutrition-related issues?
(c) How do organizations promote worker wellness through healthy eating initiatives? What benefits do they seek for their employees' well-being and organizational performance?

In the second part of the book, we take an economic perspective by providing an overview of the costs associated with malnutrition, first concerning childhood and second concerning older adults. We also consider the impact of malnutrition on healthcare costs in developed countries and the overall societal costs linked to malnutrition. Then, we present extant policies to improve the system and address malnutrition, with a focus on policies for a healthy diet and for childhood malnutrition.

Finally, we adopt an organizational perspective of diet and nutrition, which has been widely overlooked by extant literature. In this direction, we first provide a theoretical overview of corporate wellness programs and diet-related interventions. Second, we present case studies of firms introducing balanced and healthy food options for their employees and of collective catering companies offering their business clients balanced menu options and ancillary services such as nutritional counseling. The chapter benefits from the analysis of direct interviews with the firms' knowledgeable informants, which offer empirical evidence of implemented practices and key insights to the readers.

This book has been financially supported by the National Recovery and Resilience Plan (NRRP), Project title "OnFoods—Research and innovation network on food and nutrition Sustainability, Safety and Security—Working ON Foods". More specifically, this manuscript stems from the joint work of a group of researchers involved in the OnFoods Spoke 6 "Tackling malnutrition". The intellectual journey of this book has been an enriching experience: working with a multidisciplinary team allowed us to open up our own perspectives and discover new ways of looking at phenomena. We would like to thank the team of The Laboratory of Dietetics and Clinical Nutrition of the University of Pavia (Prof. Hellas Cena, Dr. Nagaia Madini, Dr. Sara Santero and Dr. Giulia Viroli) for contributing to the writing of this book. We are very grateful to Dr. Santero and Dr. Viroli for supporting us in the data collection process.

This book focuses primarily on students and researchers in nutrition and management and organization studies, but it is also informative for practitioners working in organizations, as it offers insightful case studies on practices that have already been implemented or experimented with in the workplace to improve employees' well-being through healthier dietary habits. Organizations may be inspired by the presented successful practices and may also attempt to experiment with research activities to find innovative ways to improve their workers' health through diet-related interventions. Additionally, in this book, we address policymakers by highlighting extant policies that have been implemented to address the three main forms of malnutrition involving the analyzed targets. A specific focus will be devoted to policies tackling childhood malnutrition; as children will become tomorrow's citizens, they must be given particular attention to ensure them a better future.

In the concluding chapter, we present our theoretical contribution, highlight some managerial implications, and offer some suggestions for possible future research directions.

Pavia, Italy Beatrice Re
Pavia, Italy Pietro Previtali

Acknowledgments

We are grateful to the Laboratory of Dietetics and Clinical Nutrition of the University of Pavia for the fruitful collaboration and to all the interviewees who supported us and provided insights for our research.

This research was funded under the National Recovery and Resilience Plan (NRRP), Mission 4 Component 2 Investment 1.3—Call for tender No. 341 of 15 March 2022 of the Italian Ministry of University and Research funded by the European Union—NextGenerationEU. Project code PE00000003, Concession Decree No. 1550 of 11 October 2022, adopted by the Italian Ministry of University and Research, CUP F13C22001210007, Project title "ON Foods—Research and innovation network on food and nutrition Sustainability, Safety and Security—Working ON Foods".

Contents

1 Introduction 1
N. Madini

2 Theoretical Background 5
S. Santero, G. Viroli, and H. Cena

3 Economic Impacts of Malnutrition 51
Beatrice Re, G. Viroli, S. Santero, and Pietro Previtali

4 Healthy Eating Strategies in the Workplace: An Organizational Perspective 65
Beatrice Re and Pietro Previtali

5 Concluding Remarks 81
Beatrice Re

List of Figures

Fig. 2.1 Globally attributable deaths for females and males in 2019 for major risk factors. (Source: [1]) 6
Fig. 2.2 Burden of malnutrition: the intergenerational transmission. (Source: [3]) 10
Fig. 2.3 Links between poverty and malnutrition. (Source: [37]) 12
Fig. 2.4 National prevalence of child and adolescent overweight and obesity, 1990–2021, with forecasts to 2050. (Source: [62]) 17
Fig. 2.5 Socioecological framework for understanding the dynamic interrelationships between personal and environmental factors influencing child and adolescent obesity onset. (Source: [69]) 18
Fig. 2.6 Main intervention strategies for addressing FS in autistic individuals, depending on hypersensory or hypersensory perception. (Source: [111]) 23
Fig. 2.7 U.K. Foresight obesity system map highlighting the complex existing network of factors influencing obesity. (Source: [148]) 28
Fig. 2.8 Pathophysiology of malnutrition in hospitalized adults. (Source: [157]) 32
Fig. 2.9 Malnutrition–hospitalization cycle. (Source: [182]) 34

List of Tables

Table 4.1 Comprehensive data collection. Source: Authors' own work 69
Table 4.2 Data collection Elior Hospes. (Source: Authors' own work) 71
Table 4.3 Data analysis. (Source: Re & Previtali [48]) 73

List of Boxes

Box 1 Tackling Food Insecurity in Italy. Case Study: Banco Alimentare . . . 14
Box 2 Brief Description of the Cross-Sectional Phase of the SENIOR Project, Conducted by the Laboratory of Dietetics and Clinical Nutrition (LDNC) of the University of Pavia, in Collaboration with ICS Maugeri IRCCS Hospital, Fondazione IRCCS Policlinico san Matteo and Fondazione IRCCS San Gerardo Dei Tintori di Monza . . . 30
Box 3 Malnutrition Screening Tools for Hospitalized Patients in Italy . . . 35
Box 4 Best Practices Implemented in Italy for Systematically Diagnosing and Treating Malnutrition in Hospitalized Patients: The Lombardy Case Study . . . 35

Abbreviations

B2B	Business to business
BAPEN	British Association for Parenteral and Enteral Nutrition
BIA	Bioelectrical impedance analysis
BMI	Body mass index
CW	Corporate wellness
DBM	Double burden of malnutrition
DRM	Disease-related malnutrition
DRVs	Dietary reference values
EDs	Eating disorders
ESPEN	European Society for Clinical Nutrition and Metabolism
FAO	Food and Agriculture Organization of the United Nations
FoPL	Front-of-pack labeling
FS	Food selectivity
FSMPs	Foods for special medical purposes
GLIM	Global Leadership Initiative on Malnutrition
HFSS	High-fat/salt/sugar
HR	Human resources
LDNC	Laboratory of Dietetics and Clinical Nutrition
LICs	Low-income countries
LMICs	Low- and middle-income countries
NCDs	Non-communicable diseases
ONS	Oral nutritional supplements
SES	Socioeconomic status
SSBs	Sugar-sweetened beverages
TBM	Triple burden of malnutrition
UNICEF	United Nations Children's Fund
UPFs	Ultra-processed foods
WHO	World Health Organization
WHtR	Waist-to-height ratio

Chapter 1
Introduction

N. Madini

Aim of the Book

This book aims to provide data and insights and illustrate different facets of malnutrition affecting a variety of targets: people with low socioeconomic status, children and adolescents, hospitalized patients, the elderly, and organizations' employees.

This book seeks to provide answers to the following research questions:

- How is malnutrition conceptualized in the current literature? What is the triple burden of malnutrition?
- How does malnutrition affect different target groups?
- What are the economic impacts of malnutrition and related societal costs?
- How do organizations foster employee well-being through healthy eating initiatives? What benefits do they seek for their employees' well-being and organizational performance?

We answer these questions by reviewing recent and current literature on the topic and then conducting a qualitative study based on in-depth interviews to collect data on best practices concerning diet-related initiatives that have been implemented by some virtuous firms in Italy.

Book Structure

This book is structured as follows. After this first introductory chapter illustrating the project's aims and structure, we introduce the first part of the book, which is devoted to clarifying how malnutrition is conceptualized in the extant literature. We

N. Madini (✉)
Laboratory of Dietetics and Clinical Nutrition, University of Pavia, Pavia, Italy
e-mail: nagaia.madini@unipv.it

© The Author(s) 2026

B. Re, P. Previtali (eds.), *Tackling Malnutrition in a Complex World*,
SpringerBriefs in Business, https://doi.org/10.1007/978-3-032-26438-1_1

first explain theoretically the meaning of "triple burden of malnutrition" (TBM), which is an "umbrella" term aimed to include three different facets of the phenomenon, namely undernutrition, overweight/obesity, and micronutrient deficiencies. We then focus on the double burden of malnutrition (DBM), which refers to the coexistence of undernutrition and overnutrition within the same population throughout the life course. The term 'across the life course' highlights the phenomenon whereby undernutrition during early developmental stages increases the likelihood of overnutrition in adulthood. This emphasizes the need to address both conditions simultaneously rather than separately, since a lot of countries worldwide are dealing with both challenges simultaneously. DBM is actually a global issue, affecting both high- and low-income countries, and poses a particular concern in contexts where stunting prevalence remains high [1].

We then proceed by explaining how malnutrition relates to different target groups, i.e., people with low socioeconomic status, children and adolescents, autistic individuals,[1] hospitalized patients, elderly people, and organization employees. We do so in the context of Italy, a country where the overt/risk for malnutrition involves 49% of hospitalized patients in the country and up to 69% of those in long-term care facilities [2]. Among children, the prevalence ranges from 25% in hospital settings to 30% in those with chronic conditions (ibid).

We subsequently investigate the economic impact of malnutrition, especially with respect to childhood and older adults. We pay particular attention to the impact of malnutrition on healthcare costs and to societal costs associated with malnutrition. Additionally, we offer an overview of the policies that have been implemented to tackle malnutrition, with a focus on the European context and on those policies concerning a healthy diet and childhood malnutrition.

Finally, we adopt a business and organizational perspective on the topic. We first introduce at a theoretical level the concept of corporate wellness programs in organizations and the emerging topic of nutrition and mental health in the workplace. We subsequently offer case studies of for-profit firms implementing activities to address malnutrition in different forms, either for their employees and/or for external companies. Additionally, we present a research project aimed at exploring how workplace nutrition can be adapted to meet the needs of autistic individuals. We conclude the book by highlighting our theoretical contribution and managerial implications and by proposing future research directions.

[1] Throughout the entire manuscript we will adopt the identity first language because it is the one preferred by the autistic community, see: https://autisticadvocacy.org/about-asan/identity-first-language/.

References

1. FAO, IFAD, UNICEF, WFP, WHO (2024). The state of food security and nutrition in the world 2024—financing to end hunger, food insecurity and malnutrition in all its forms. Rome
2. Pradelli L, Zaniolo O, Sanfilippo A, Lezo A, Riso S, Zanetti M (2023) Prevalence and economic cost of malnutrition in Italy: a systematic review and metanalysis from the Italian Society of Artificial Nutrition and Metabolism (SINPE). Nutrition:111943

Open Access This chapter is licensed under the terms of the Creative Commons Attribution-NonCommercial-NoDerivatives 4.0 International License (http://creativecommons.org/licenses/by-nc-nd/4.0/), which permits any noncommercial use, sharing, distribution and reproduction in any medium or format, as long as you give appropriate credit to the original author(s) and the source, provide a link to the Creative Commons license and indicate if you modified the licensed material. You do not have permission under this license to share adapted material derived from this chapter or parts of it.

The images or other third party material in this chapter are included in the chapter's Creative Commons license, unless indicated otherwise in a credit line to the material. If material is not included in the chapter's Creative Commons license and your intended use is not permitted by statutory regulation or exceeds the permitted use, you will need to obtain permission directly from the copyright holder.

Chapter 2
Theoretical Background

S. Santero, G. Viroli, and H. Cena

Triple Burden of Malnutrition: Key Concepts, Data, and Insights

Key Points

- **Malnutrition** can manifest in all individuals, in all forms, and includes undernutrition (wasting, stunting, underweight), overnutrition (overweight, obesity, and resulting diet-related noncommunicable diseases) and/or inadequate vitamins or minerals.
- The **TBM** is the concurrent manifestation of undernutrition, overnutrition, and/or micronutrient deficiencies within individuals, households, and populations, persisting across the life course: 194 million children under five years are living with stunting or wasting while 37 million with overweight or obesity. However, during the last years, while underweight is decreasing, overweight and obesity are increasing in most age groups.
- Malnutrition has deep and **long-term effects** on development, health, economies, and societies, impacting not only individuals and their families but also entire communities and nations.

S. Santero (✉) · G. Viroli
Laboratory of Dietetics and Clinical Nutrition, University of Pavia, Pavia, Italy
e-mail: sara.santero@unipv.it; giulia.viroli@unipv.it

H. Cena
Laboratory of Dietetics and Clinical Nutrition, University of Pavia, Pavia, Italy

Istituti Clinici Scientifici Maugeri IRCCS, Unit of Clinical Nutrition, Pavia, Italy
e-mail: hellas.cena@unipv.it

© The Author(s) 2026

B. Re, P. Previtali (eds.), *Tackling Malnutrition in a Complex World*,
SpringerBriefs in Business, https://doi.org/10.1007/978-3-032-26438-1_2

Unbalanced diets are among the top global risk factors for all-cause death and increase the risk of several diseases, such as diabetes, obesity, cardiovascular diseases, certain types of cancer, loss of disability-adjusted life years, poor quality of life and premature death [1]. Adequate nutrition not only provides health benefits but also economic and social advantages from the individual to families and entire communities, regions, and countries of current and future generations [2].

Figure 2.1 shows globally attributable deaths for females and males in 2019 for major risk factors.

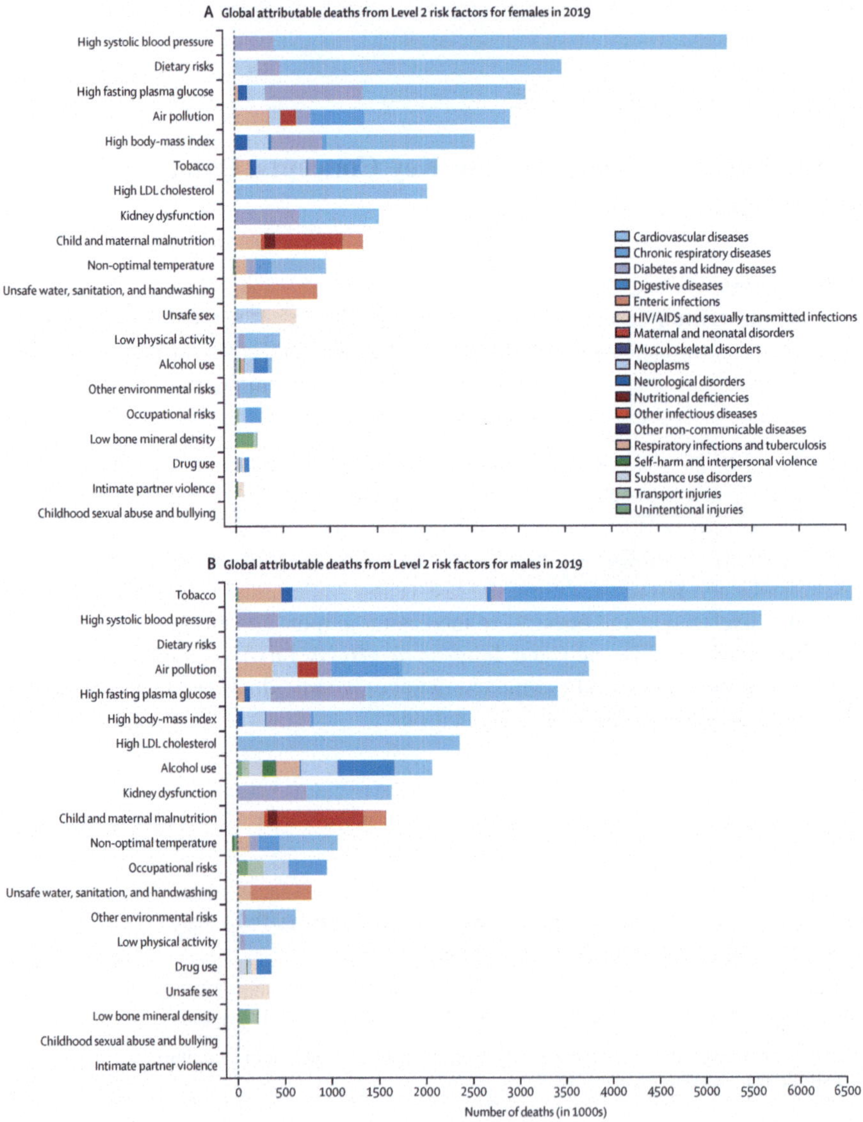

Fig. 2.1 Globally attributable deaths for females and males in 2019 for major risk factors. (Source: [1])

Historically, undernutrition and overnutrition have been viewed and approached as two distinct phenomena that require different interventions. However, it is currently clear that these forms of malnutrition are not isolated phenomena but are rather closely interrelated. This concept is called the **DBM**, which refers to the "coexistence of undernutrition along with overweight, obesity or diet-related non-communicable diseases (NCDs), within individuals, households and populations, and across the life-course" [3]. This new paradigm first emerged in 1992, highlighting that separating undernutrition from overweight/obesity was no longer useful since most countries, especially Low-and Middle-Income Countries (LMICs), were dealing with both problems simultaneously. The term DBM has been replaced by the term TBM, which includes micronutrient deficiencies as well as undernutrition and overnutrition [3]. However, studying **micronutrient deficiencies** is more complicated and resource-intensive than measuring undernutrition or overnutrition. It often requires blood tests or other lab analyses, which can be expensive and difficult to carry out on a large scale. Because of this, these deficiencies are less commonly studied, and there is less evidence available about this aspect of the **TBM**. Consequently, while recognizing that micronutrient deficiencies is a key issue to be addressed, as underlined in the subparagraph below (Section "Micronutrient Deficiency"), we then opt to focus solely on malnutrition due to deficiency and excess (Section "Between Deficiency and Excess: Driving Factors of the Double Burden").

The Three Forms of Malnutrition: Undernutrition, Micronutrient Deficiency, and Overnutrition

The World Health Organization (WHO) referred to the term **malnutrition** to describe "nutrition" in three broad categories: undernutrition, micronutrient-related malnutrition, and overweight or obesity and diet-related NCDs [4, 5]. We provide an overview of each of these forms.

Undernutrition

Undernutrition encompasses 3 forms: wasting, stunting, and underweight. **Wasting**, or acute malnutrition, is characterized by a low weight-for-height ratio. It is typically associated with recent and severe weight loss due to inadequate food intake or disease, often triggered by crises such as conflict or climate events. Wasting can range in severity from mild (−1 to −2 Z-scores) to severe (below −3 Z-scores), with the latter posing a significant risk of mortality in children and requiring immediate intervention [6]. **Stunting** is characterized by low length/height for age. It reflects an indication of prolonged nutritional deficiencies often rooted in poverty, poor infant feeding practices, and recurrent illness. This condition can lead to irreversible impairments in both physical and cognitive development (ibid). **Underweight** is defined as low weight-for-age (ibid). It may result from both acute and chronic undernutrition and can also present varying degrees of severity.

According to The State of Food Security and Nutrition in the World 2024 [7], global stunting among children under five declined from 26.3% in 2012 to 22.3% in 2022, but the global prevalence of wasting in children under 5 years of age has remained relatively unchanged, with a slight decline from 7.5% in 2012 to 6.8% in 2022, mostly in low-income countries (LICs) and LMICs.

Micronutrient Deficiency

Micronutrient deficiency is the lack of vitamins and minerals required for the production of essential components needed for growth, development and survival. It affects more than 2 billion people worldwide, with dramatic inequities that particularly affect vulnerable populations such as pregnant women, people with illnesses and older people. Under the international organization's spotlight, prioritization has been given to anemia, vitamin A, and iodine deficiency for worldwide prevention, control and elimination [8]. **Anemia** is a complex health condition commonly caused by iron deficiency, although a lack of folates and vitamin B12 are also important causes. It affects almost 30% of the population, particularly young children and women of reproductive age (30%) and pregnant women (37%). Consequences can vary from fatigue to dizziness, maternal deaths during childbirth and shortness of breath [9, 10]. **Vitamin A** deficiency is a severe public health problem that is estimated to cause blindness in 250 million children, and half of them die within one year of becoming blind, making this vitamin the leading cause of preventable blindness [11]. **Iodine** deficiency impairs thyroid function, leading to developmental and cognitive damage, especially in infants born to iodine-deficient mothers. It remains the most common cause of preventable brain damage globally, notably in LICs [12].

Overweight and Obesity

Overweight and **obesity**, defined as excessive fat accumulation, are classified using body mass index (BMI), with thresholds of $\geq$25 kg/m^2 for overweight and $\geq$30 kg/m^2 for obesity in adults [13]. However, recent research highlights a shift from BMI alone to a more comprehensive clinical staging approach. These methods consider adipose tissue function, fat distribution, metabolic complications, and obesity-related impairments, providing a more accurate diagnosis of the disease's heterogeneity [14, 15]. Among children and adolescents, the use of growth charts is necessary for assessment (see Section "Target II: Malnutrition in the Developmental Age").

Obesity is a multifactorial chronic disease influenced by individual, social and environmental factors. It is considered a twenty-first-century epidemic and a key risk factor for NCDs such as type 2 diabetes, cardiovascular diseases, metabolic syndrome, hypertension, respiratory diseases, and some types of cancer, among others [13, 16].

3 billion people worldwide are living with overweight or obesity, including about 880 million with obesity. The worldwide prevalence has more than tripled between 1975 and 2022, with alarming spikes in LMICs [17]. Childhood obesity has also emerged, increasing from just 4% in 1975 to almost 20% in 2022, affecting 159 million children in 2022 [7].

Between Deficiency and Excess: Driving Factors of the Double Burden

Since the last 30 years, DBM has continued to rise and change in form. Popkin et al. [18] reported that, at a household level, the increase of the DBM is driven by the rapid increases in overweight and obesity (especially in women), combined with a slower decline in undernutrition. This phenomenon is especially prevalent in LMICs.

To explain why, despite economic growth, improvements in food security, and better access to healthcare, overnutrition and micronutrient deficiencies have become the dominant forms of malnutrition, it is necessary to consider the complex interplay of biological, social, and environmental factors. Hereafter, we briefly address the causes of DBM, but we recommend looking into the academic literature for a more in-depth discussion [3, 18–22].

Biological and health factors help explain the occurrence of the DBM. Nutritional deprivation during fetal development can alter gene expression and metabolic function, increasing the risk of obesity and NCDs later in life [23, 24]. As illustrated in Fig. 2.2, maternal undernutrition, smoking, and exposure to endocrine disruptors, as well maternal overweight and obesity contribute to this intrauterine reprogramming, which can also affect health in later life.

Beyond biology, **early-life practices** such as delayed or absent breastfeeding and inadequate weaning significantly impair nutritional status, increasing susceptibility to infections and chronic diseases. Moreover, the transformation of the **global food environment**, driven by industrialization, urbanization increase the availability of convenient unhealthy foods in homes, schools and workplaces [25–28]. This shift has also led to a decline in green spaces, physical activity and increased screen time [20]. Finally, non-resilient food supply chains, economic crises, conflicts, and **climate crises** further increase malnutrition risk [29].

The combination of these three elements, namely the cluster of twenty-first-century pandemics (climate change, undernutrition, and overnutrition), defined as

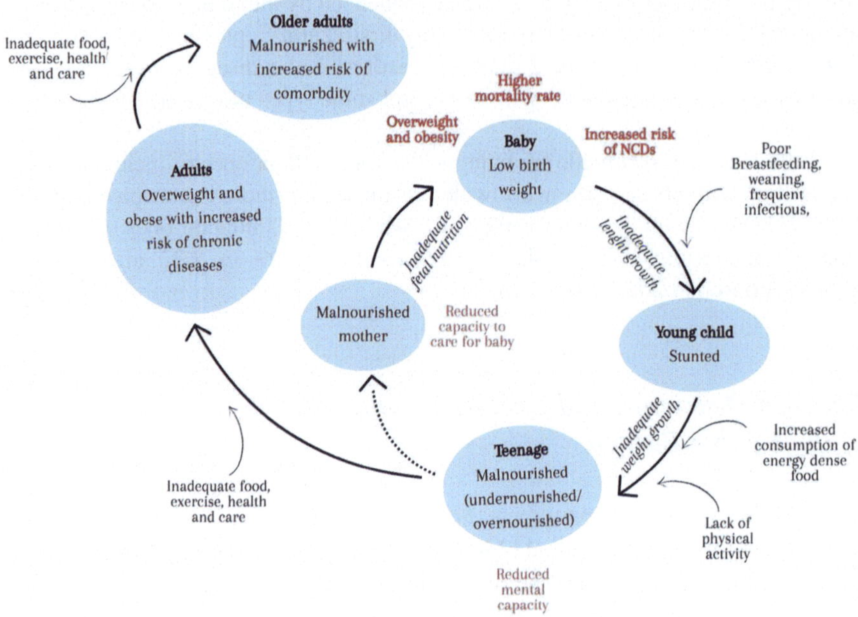

Fig. 2.2 Burden of malnutrition: the intergenerational transmission. (Source: [3])

the "**Global Syndemic**" by the Lancet Commission, calls for a transformation of food systems. Sustainable healthy diets should be central to this strategy, addressing all forms of malnutrition while safeguarding planetary health [29].

An Overview of the Italian Scenario

Nowadays, the food habits of the Italian population are far from being in line with Italian food-based dietary guidelines [30]. For instance, the latest data on dietary patterns of the Italian population reveal low consumption of fruits, vegetables, legumes, whole grains, and fish [31]. Additionally, in 2022–2023, 32.7% of adults were overweight, 10.4% with obesity, and almost half of the population carried excess weight, particularly in southern regions [32]. Moreover, anemia affects 13.6% of women aged 15–49 years [33].

However, nutritional status and malnutrition in Italy are not uniform across the population. Certain groups, such as those living in poverty, older adults, children, and adolescents, are particularly vulnerable to developing malnutrition. These groups are precisely those addressed in this book—as explained in the introduction—and we offer an overview of them in the following subparagraphs.

Target I: Malnutrition and Poverty

Key Points

- **Poverty and malnutrition are deeply linked**, leading to both undernutrition and, paradoxically, rising obesity.
- **Definitions:**
  - **Poverty:** deprivation of basic amenities that restricts individuals from leading a healthy life, including but not limited to healthcare, education, hygiene and diet.
  - **Hunger:** uncomfortable or painful physical sensation caused by involuntary insufficient consumption of dietary energy.
  - **Hidden hunger:** subclinical nutrient deficiencies (e.g., iron, folic acid, vitamin A), but does not have overt clinical signs of undernutrition.
  - **Food security:** all individuals, always, have physical, social, and economic access to sufficient, safe, and nutritious food to meet their dietary needs and food preferences for an active and healthy life.

    Food insecurity: state where there is limited, inadequate, or unreliable availability or access to obtain nutritionally sufficient and safe foods in socially acceptable ways. It can be **moderate** (decrease the quality and/or quantity of the food consumed) or **severe** (regularly runs out of food and, in the most extreme cases, may go a day, or several days, without eating, often resulting in hunger).
- **The rising cost of healthy diets**, worsened by global crises and climate change, is pushing billions into food insecurity, with serious health consequences across all age groups and regions.

Malnutrition is a global phenomenon affecting people living in poverty to a greater extent. According to the World Bank, about 8.5% of the global population lived in extreme poverty in 2024, which means they lived on less than $2.15 per person per day. Compared to nearly 36% in 1990 who lived under the poverty line, the proportion of the world's population living in extreme poverty has declined by more than half. The economic barrier to food may contribute to the onset of undernutrition by not fulfilling an individual's nutritional requirements, resulting in a deficiency of energy and strength necessary for daily activities, impairing work capacity and productivity and, in the worst case, experiencing hunger. The Food and Agriculture Organization of the United Nations (FAO) defines hunger as an uncomfortable or painful physical sensation caused by insufficient consumption of dietary energy. For decades, the prevalence of undernourishment has been utilized as an indicator of hunger worldwide. Despite progress, it is estimated that more than 600 million people will still face hunger in 2030 [34].

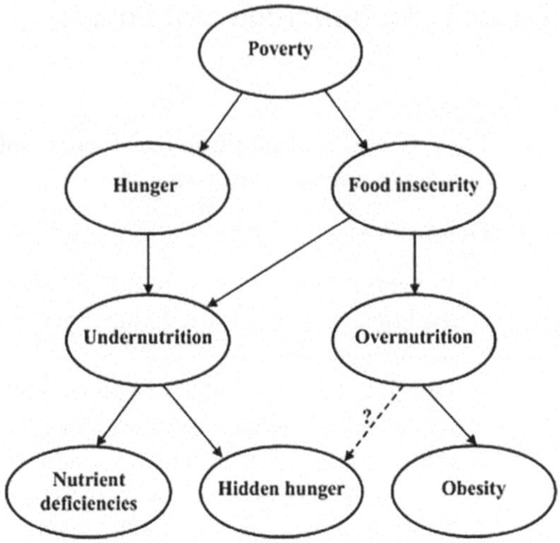

Fig. 2.3 Links between poverty and malnutrition. (Source: [37])

However, poverty is a complex and multifaceted condition that goes beyond income [35]. It embraces a holistic deprivation of basic amenities that restricts individuals from leading a fulfilling life, including, but not limited to, assets, living standards, education, sanitation and hygiene, health, and nutrition. Understanding the link between malnutrition and poverty requires considering social determinants of health: conditions and circumstances in which people live, learn, work, and even play. People living in poverty face greater environmental risks, limited coping capacity, reduced access to quality healthcare, and barriers to obtaining nutritious foods, all of which contribute to malnutrition [35, 36].

As illustrated in Fig. 2.3, poverty can manifest not only with undernutrition but also with overweight and obesity. Regions such as Africa, Latin America, and the Caribbean, which face high poverty rates, are experiencing some of the highest global rates of obesity [38]. The obesity epidemic has been characterized by a wave-shaped trajectory that differs by SES group [39, 40]: in the early stages, individuals in higher SES groups tend to have higher obesity rates; however, as countries transition to later stages, obesity increasingly affects lower SES groups, reversing the initial trend. Some evidence from Europe and the United States supported this model by showing that individuals from lower SES backgrounds were approximately twice as likely to live with obesity [41]. In line with this, an analysis across 37 developing countries indicates that, in low-income economies, lower SES was associated with reduced obesity incidence, but in upper-middle-income developing countries lower SES became a consistent risk factor for obesity [42, 43].

As a result, both undernutrition and overnutrition may arise from living in poverty. Poor diets often result in insufficient micronutrient intake or hidden hunger, which, during pregnancy, can impair a child's health, as well as future educational and economic outcomes later in life [23, 24]. Consequently, both malnutrition and poverty are characterized by intergenerational transmission, in which both poverty and malnutrition fuel and reinforce each other. Breaking this cycle, particularly by reducing poverty among women, is therefore essential [44]. Like the DBM, the vicious cycle of poverty and malnutrition can affect not only individuals but also families, communities and countries. Persistent poverty and malnutrition threaten community and national economic stability, whereas improvements in safety nets and access to services for the poor have been shown to support national growth. Where poverty persists, even in wealthy nations, hunger remains a problem [45].

Food Insecurity

Hunger can also be, but not necessarily, a consequence of food insecurity. According to the Rome Declaration on World Food Security 1996, food insecurity is as a state in which "all individuals, always, have physical, social, and economic access to sufficient, safe, and nutritious food to meet their dietary needs and food preferences for an active and healthy life" [7, 44]. It can manifest at varying degrees of severity: severe food insecurity occurs when an individual regularly runs out of food and, in the most extreme cases, may go a day, or several days, without eating. Moderate food insecurity forces individuals to decrease the quality and/or quantity of the food. The Integrated Food Security Phase Classification [46] differentiates five distinct phases of the severity of acute food insecurity: (1) minimal/none, (2) expressed, (3) crisis, (4) emergency, and (5) catastrophe/famine. This classification provides relevant information to decision-makers that focuses on short-term objectives to prevent or mitigate life threatening food insecurity.

Globally, nearly 29% of the population (2.33 billion people) faces moderate or severe food insecurity, with the highest prevalence in Africa (58%) and other countries (Latin America, the Caribbean, Asia and Oceania). In Europe, North America, Australia and New Zealand, the prevalence is much lower (8.9%). Women and rural populations are generally more affected, except in HICs where urban populations face higher food insecurity. Food insecurity is driven by political instability, economic inequality, conflict, and discrimination. Climate change worsens these factors by disrupting the agriculture system through extreme weather, rising temperatures, and water scarcity, reducing food production and availability [29, 47].

Food insecurity, especially in its moderate form, plays a critical role in the DBM, where both undernutrition and overnutrition coexist [37]. In such situations, individuals may be surrounded by a variety of food options, but their access is limited primarily to low-cost, energy-dense foods high in carbohydrates, saturated fats, sugars, and sodium rather than nutritionally rich and diverse food choices. Thus, increasing the risk of NCDs in adults [48, 49] and older adults [50, 51]. Moreover, living in poor urban areas, where healthy food and safe spaces for exercise are scarce, further exacerbates the risk of obesity, while inconsistent food access and deprivation episodes can trigger stress and physiological responses that promote adipose tissue accumulation. This phenomenon has also been observed in children [34, 52, 53].

Global food price increases have limited access to healthy diets [54]. Between 2020 and 2022, the cost rose from 3.35 to 3.96 PPP dollars, leaving over 2.8 billion people unable to afford a healthy diet, especially in Latin America and Asia [7]. In Europe, food insecurity affects at least 10% of urban populations in 16 of 24 countries [55].

The Italian Scenario

According to the Italian National Institute of Statistics [56], in 2023, almost 5.7 million individuals (9.7%) lived in absolute poverty, meaning that they lived with a monthly income below the minimum necessary expenditure. Poverty was higher among families with at least one foreign member (30.4%) than among Italian families (6.3%). Geographically, the South recorded the highest rates (over 859,000 families, 10.2%), followed by the Northwest (8.0%, 585,000 families) and the Northeast (7.9%, 413,000 families). The Centre had the lowest levels (6.7%, 360,000 families). Among minors, the poverty rate reached 13.8% (1.3 million children), the highest since 2014, while it was 11.8% in the 18–34 age group and 6.2% among older people. Regarding food insecurity, 1.5% of Italians experienced moderate or severe food insecurity. As for poverty, the prevalence was highest in the South (2.7%), compared to the North (0.8%) and the Centre (1%). Food insecurity was more common in urban areas (1.8%) and medium-sized cities (1.2%) than in rural areas (1%). Nearly 6% of children and adolescents under 16 years of age showed signs of food deprivation, especially in single-parent households [57].

In 2022, 3.1 million people required public assistance to eat soup kitchens or receive food parcels, including over 630,000 children under 15 years old and 356,000 elderly people. More than half of the recipients were women, and 72.3% had children [58].

Box 1 Tackling Food Insecurity in Italy. Case Study: Banco Alimentare

TACKLING FOOD INSECURITY IN ITALY

Case Study: Banco Alimentare

Banco Alimentare was born in Arizona in 1967 from an idea of John van Hengel, who was the first to start distributing to the most deprived the surplus food from shops and restaurants. In 1989 Banco Alimentare also settled in Italy with the establishment of the Foundation in Muggiò (North of Milan).

Currently, it includes the Foundation and 21 Banco Alimentare Organizations across Italy. Banco Alimentare's vision is to contribute to alleviating the problem of hunger, marginalization and poverty, as well as promote the fight against food waste for social purposes. To do so, it coordinates and organizes the recovery of surpluses from the food supply chain, through Banco Alimentare Organisations which, in turn, distribute the food free of charge to the Territorial Partner Organizations that welcome people and families in need. The food is recovered from the entire agri-food chain (agricultural surpluses), industry, large-scale distribution platforms, supermarkets, and collective catering.

Through an in-depth 2 h interview that we conducted in July 2024 with Dr. Marco Lucchini, Secretary General, we investigated the Foundation's goals, its main targets, and the Foundation's activities to tackle malnutrition in people with economic difficulties.

First, Dr. Lucchini underlined that Banco Alimentare primarily deals with Territorial Partner Organizations (not-for-profit organizations). Their shared aim is to welcome and support the most deprived facing issues in stable access to food. In his words "our role is subsidiary to the organizations as regards the food aspect so that they can best carry out their specific purpose, which is to promote in various ways the inclusion of people in different difficult situations. Our support in terms of food aid means that these organizations have more time and resources to dedicate to their core projects. We work with the entire food chain, acting as a bridge between for-profit and not-for-profit organizations."

The guiding principles of the Foundation concern the provision of food in a sufficient, healthy, nutritious way. By following the requests provided by Territorial Partner Organizations, it focuses on social and cultural acceptability of its activities. As reported by our interviewee:

> What makes us peculiar is that we combine food recovery and distribution with the social dimension: the recovery and transfer of surplus food occurs free of charge to organizations and then consequently to the end beneficiaries.

Moreover, whereas in the past there was mainly a focus on the ethnic and religious aspects linked to food, the sensitivity has progressively been extended to a broader spectrum of food needs, like those concerning veganism and celiac disease. The respect for food preferences and intolerances is indeed another key trait of the organization.

The commitment to support people in need emerges from an interesting project carried on by the organization in 2016–2017 in collaboration with Kellogg's, which is called "Evviva la colazione" (Breakfasts for Better Days™). The project initially involved children because it was born from the awareness that children living in absolute poverty often skip one of the main meals of the day, breakfast. Indeed, some research [59] reports that 8% of Italian children skip breakfast, while 31% of them have an unbalanced breakfast (i.e., without the right amount of carbohydrates or proteins). Additionally, a recent study [60] highlights that 1 in 10 children in Italy faces socio-economic challenges that prevent them from accessing sufficient and nutritious food on a daily basis.

The Breakfasts for Better Days™ project aimed to help reduce these inequalities by collecting products for breakfast and donating them to the families of the 140,000 children helped throughout Italy by the Food Bank Network. On the other hand, it was also aimed at raising awareness among families about proper nutrition. Due to the success of the project, it was extended in 2018 to the elderly as well with Kellogg's, Banco Alimentare managed to help over 10 thousand Italian older adults with a low SES. A nutritious breakfast was personally delivered to their homes. In 2019, the collaboration led to the donation of over 155,000 kg of products for the first meal of the day, for a total of 445,000 breakfasts. This meant that 42 breakfasts have been delivered to each elderly person involved in the project.

Summing up, Kellogg's company donates food to the European Food Bank Network and to not-for-profit associations. Specifically, donations are redistributed in 20 European countries, including Italy, and other nations such as Mexico, Canada and South Africa. According to data reported on the Kellogg's website, in 2020 the company donated more than 150,000 kg of food in Italy to organizations, providing more than five million breakfasts and snacks to local communities and people with a low SES. This fruitful collaboration shows that for-profit organizations and not-for-profit ones can develop significant synergies to give a contribution to tackle food insecurity.

Target II: Malnutrition in the Developmental Age

Key Points

- **Childhood obesity is rapidly increasing**: 244% rise since 1990 and projections showing a further 121% increase by 2050, making it one of the most serious public health issues of our time.
- **Obesity in children is influenced by multiple factors,** including biology, lifestyle, family environment, and socioeconomic conditions, which requires a broad, socioecological approach for prevention and intervention.
- **The health consequences of childhood obesity are extensive**, affecting multiple body systems and increasing the risk of chronic diseases, disability, and premature death, along with significant economic and societal costs.
- **Childhood obesity is linked to mental health**, including depression, anxiety, low self-esteem, and social stigma, all of which can worsen health outcomes and quality of life, creating a vicious cycle unless addressed with integrated, compassionate care. Psychological disorders seem to have a **bidirectional relationship** with obesity, where each condition can both influence and worsen the other.

By the 2050s, the number of children under 18 in the world is expected to remain close to the current number, approximately 2.3 billion [61]. Malnutrition persists and although stunting prevalence is declining [7], a sharp rise in childhood overweight and obesity is observed. Indeed, their prevalence increased by 244% between 1990 and 2021, and is projected to rise by another 121% by 2050 (see Fig. 2.4) [62]. Thus, pediatric obesity has been described as one of the most serious public health challenges of the twenty-first century since it increases the risk of developing type

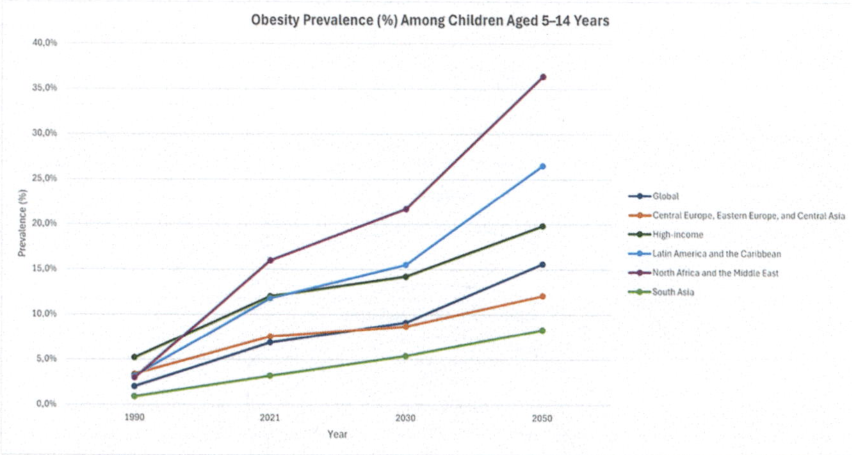

Fig. 2.4 National prevalence of child and adolescent overweight and obesity, 1990–2021, with forecasts to 2050. (Source: [62])

2 diabetes and cardiovascular diseases, premature death, and long-term disability even during childhood [62].

Childhood obesity is diagnosed using BMI relative to age- and sex-specific growth charts. WHO standards 0–5 years and 5–19 years are commonly used, with overweight and obesity defined by SD cutoffs [6, 63–65]. The Centers for Disease Control and Prevention (CDC) charts (2–20 years) [66] and International Obesity Task Force (IOTF) tables (2–18 years) [67] are also applied. The diagnosis often has to be complemented with anthropometric measurements, lifestyle assessment, medical history, and family background [68].

Obesity is a complex and multifactorial disease in which the energy imbalance between energy intake and energy expenditure is the ultimate manifestation of fat accumulation. Similarly to DBM, it is influenced by biological, lifestyle, environmental, and socio-economic factors within family and cultural contexts (Fig. 2.5) [69]. Understanding these risk factors is essential for effective prevention, early intervention, and the development of comprehensive **public health strategies** [48, 68–70].

The **consequences** of the onset of childhood obesity span multiple organ systems and affect physical, psychological, and societal health. Children and adolescents with obesity are at significantly higher risk for several diseases, including hypertension, dyslipidemia, insulin resistance, type 2 diabetes mellitus, nonalcoholic fatty liver disease, reduced pulmonary function, obstructive sleep apnea syndrome, chronic joint pain, skeletal deformities, early puberty, menstrual irregularities and polycystic ovary syndrome (PCOS) [68, 69].

Given the broad negative effects of obesity on health, the long-term consequences of obesity are particularly concerning since children with obesity will likely be adults with obesity, who will be at increased risk of developing NCDs. Moreover, the burden of disease often results in reduced workforce participation, productivity loss, and increased healthcare utilization, highlighting the economic and societal costs of early-onset obesity.

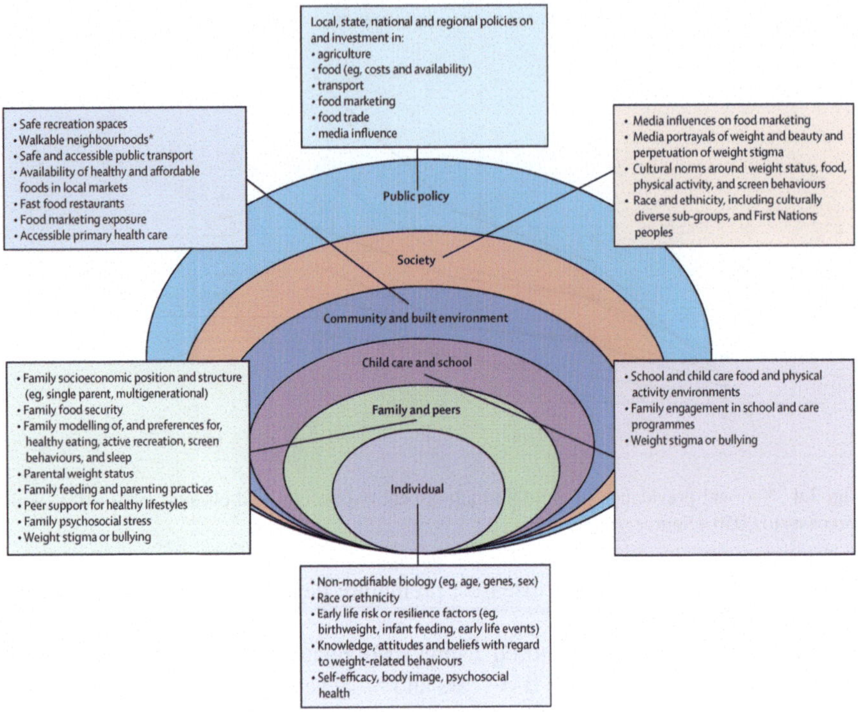

Fig. 2.5 Socioecological framework for understanding the dynamic interrelationships between personal and environmental factors influencing child and adolescent obesity onset. (Source: [69])

Psychological Impacts of Obesity in Children

In addition to physical health consequences, obesity can pose significant psychological effects. Children and adolescents with overweight or obesity face significantly greater risks of psychological comorbidities than their healthy-weight peers do, including depression, anxiety [48, 71–74], reduced health-related quality of life and low self-esteem [75].

Moreover, obesity is considered to be one of the "most stigmatizing and least socially acceptable conditions in childhood" [76]. Obesity stigma is widespread in a variety of ways, from parents to peers and even health professionals. For children with obesity, this stigma increases the risk of social isolation, internalized weight bias, reduced physical activity, avoidance of healthcare services, further weight gain, and decreased motivation to adopt healthier behaviors [77]. Conversely, psychiatric disorders may also contribute to the development of obesity in vulnerable individuals, and obesity exacerbates those vulnerabilities. Psychiatric conditions such as depression [68, 72], autism (see Section "Target III—Malnutrition in Autistic Individuals") and attention-deficit hyperactivity disorder (ADHD) [78] seem to have a bidirectional association with obesity.

Together, these studies underscore the urgent need to address the psychological consequences of obesity in youth. A comprehensive, stigma-sensitive approach that incorporates mental health care alongside weight management is essential for fostering long-term health and well-being in children and adolescents affected by obesity.

Italian Scenario

In Italy, according to the 2023 data published by OKkio alla SALUTE, the national surveillance system part of the WHO European Region's Childhood Obesity Surveillance Initiative [4, 5], 19% of children aged 8–9 years have overweight, and 9.8% have obesity, including 2.6% with severe obesity. Despite a decrease in overweight rates, obesity has remained stable, with a slight increase reported in 2023. A persistent North-South geographical gradient remains, with higher rates among children in southern Italy. The lifestyle of Italian children, characterized by skipping breakfast, consuming insufficient fruits, vegetables, and legumes, while eating excessive sweet snacks, and spending long hours in front of screens, contributes to this high prevalence of overnutrition (ibid).

Target III—Malnutrition in Autistic Individuals

Key Points

- Autism is conceptualized as a **spectrum condition** that includes a range of presentations formerly categorized as separate disorders, a shift that reflects both changes in diagnostic criteria over time and the diversity of theories and causal explanations proposed for autistic traits. Its prevalence is rising worldwide;
- Autistic individuals frequently experience a hypo- or hypersensory perception, which significantly impacts food choices and eating behaviors. Indeed, **food selectivity (FS),** a condition characterized by a limitation in the variety of foods accepted with a sensory aversion to specific textures, temperatures, flavors, colors and smells, is common in autistic people (21–76%);
- The prevalence of overweight and obesity in autistic individuals is significantly greater than that in the neurotypical population. Autistic individuals tend to follow unhealthy dietary patterns, rich in ultra-processed foods, predisposing to the development of overweight and obesity;
- Effective programs to tackle malnutrition in this population should be structured to counter FS, encompassing behavioral and educational aspects, and consider the presence of hypo or hypersensory perception. Furthermore, they should involve the caregivers and be performed by a multidisciplinary team.

Autism is conceptualized as a **spectrum condition** that includes a range of presentations formerly categorized as separate disorders (autism, Asperger's syndrome, and pervasive developmental disorders not otherwise specified) [79]. This shift reflects both changes in diagnostic criteria over time and the diversity of theories and causal explanations proposed for autistic traits. Symptoms associated with autism are typically observable from the early developmental period and influence social participation, daily activities, or other important areas of functioning [79].

In recent years, an increase in the prevalence of autism has been reported [80], which is linked both to changes in diagnostic criteria and to increased scientific evidence and knowledge of the condition among the general population. Recent American data indicated a prevalence of 1 autistic child out of 36 children [81], whereas in Europe, this prevalence ranged between 0.38% and 1.55% of the total population [82]. In Italy, data collected between 2015 and 2017 by the Istituto Superiore di Sanità and the Ministry of Health reported a prevalence of 1 in 77 children [83]. Notably, there is a well-known increase in the prevalence of autism in males, with a ratio ranging from 4:1 according to American data [84] to 3:1 according to Italian data [83].

Food Selectivity in Autistic Individuals

This vulnerable population is frequently characterized by alterations (i.e., hypo- or hyperperception) in sensory perception [85] concerning taste, smell, sight and/or touch, which certainly have an impact on eating behavior. Sensory sensitivity to food texture and taste is frequently reported and is approximately 2–10 times greater in autistic children than in control individuals [86].

Individuals with sensory hypersensitivity may feel 'overwhelmed' by the surrounding world, leading to rejection of and aversion to certain foods according to the individual's sensory hyperreactivity characteristics [87]. The person might, for example, perceive colors as bright or texture as disgusting or smell as too intense or taste as excessively bitter. On the other hand, individuals with sensory hyposensitivity might receive little stimulation from the surrounding world, potentially leading them to consume an excessive amount of food, especially savory food. Therefore, they could seek out sweet, salty, or spicy foods to reach an adequate threshold of sensory stimulation to the point of ingesting inedible objects, an eating disorder known as 'pica' [88, 89].

Considering these altered sensory perceptions, the high prevalence of food selectivity (FS) in autistic people and an intellectual ability level in the normal range, 21–76% [86], is not surprising [90]. FS, also known as "peaky" or "fussy" eating, is a condition characterized by a limitation in the variety of foods accepted, with a sensory aversion to specific textures, temperatures, flavors, colors and smells [89, 91]. Nonetheless, there is a lack of an international consensus for its definition [92]. FS often results in the development of strong preferences for specific foods and even brands, vaunting easily recognizable packaging and distinct organoleptic characteristics [89]. Such foods provide an element of continuity and 'predictability' (sameness) in a highly variable food landscape. For example, a specific brand of packaged potato crisps will always look and taste the same, whereas an unpackaged product (i.e., a banana) might be very ripe, ripe or unripe, thus being 'unpredictable' [93]. Given this background, the pronounced preference for ultra-processed foods (UPFs) among autistic individuals is not unexpected. UPFs, such as chicken nuggets, fries and packaged snacks, are highly processed, energy-dense foods that are high in simple sugars, saturated fats, and salt and low in micronutrients [94]. Moreover, this population tends to dislike fruits and vegetables because of their great variety of organoleptic characteristics, texture and aesthetic variability [93]. With respect to protein sources, there is generally a good acceptance rate, except for legumes and fish. The possible presence of bones and the characteristic strong smell/taste might be the causes of the reduced fish acceptance rate [93]. Furthermore, autistic individuals are known to prefer refined cereals over whole grains and to consume few fruits, vegetables, and legumes. This population generally shows a marked propensity toward sugary beverages at the expense of water consumption [93]. Notably, FS often persists from childhood into adolescence [95] and adulthood [96], underscoring the importance of paying attention to the nutritional needs of autistic individuals.

Nutritional Status in Autistic Individuals

The co-presence of FS in autistic individuals exposes them to excessive intake of UPFs, energy drinks and refined carbohydrates, while the consumption of fruit, vegetables, legumes and whole grains is low. An unhealthy diet (a Western diet) with a reduced dietary variety causes excessive intake of sodium, sugar, saturated fatty acids and trans-fatty acids as well as a reduced intake of fiber, vitamins and minerals. Thus, it predisposes individuals to the development of overweight and obesity with associated metabolic comorbidities [97–99] and micronutrient deficiencies [100]. Indeed, significantly lower adherence to the Mediterranean has been observed in autistic children than in healthy controls without autism [101]. In addition to these two forms of malnutrition, which are most common in this population group, autistic individuals may also develop undernutrition-related malnutrition (underweight) [97]. Finally, a high prevalence of eating disorders, especially anorexia nervosa, has been reported in this population [102, 103].

Notably, the prevalence of overweight and obesity in this 'vulnerable' population is significantly greater than that in the neurotypical population [97]. When stratified by the severity of autistic traits, obesity prevalence appears higher among autistic individuals with greater support needs than among those requiring lower levels of assistance [104]. Overweight and obesity also increase the risk of developing metabolic comorbidities known to be associated with excess weight, including hyperglycemia, type II diabetes mellitus, dyslipidemia, and hypertension, even in this population. As childhood obesity often persists into adolescence and adulthood and worsens the disability status and quality of life of these 'vulnerable' individuals [105], prevention of unhealthy weight gain from childhood in autistic individuals is crucial. The risk factors for obesity in this population group are a high rate of sedentary behavior, motor disorders, unhealthy eating patterns, excessive screen time, sleep disorders, and drugs used to control psychological and behavioral symptoms [105].

Despite the high prevalence of several micronutrient deficiencies reported in this population, there is no consensus in the literature on which micronutrient deficiencies are most frequently observed among these individuals. However, evidence seems to suggest an increased risk of calcium, phosphorus, selenium, vitamin D, vitamin B1, vitamin B2, vitamin B12, and vitamin E deficiency [100]. Furthermore, reports of food-related diseases, which are currently rare among the general population, are not uncommon in children with severe FS. These diseases include scurvy from vitamin C deficiency [106], xerophthalmia from vitamin A deficiency [107], optic neuropathy from vitamin B12 deficiency [108], and pellagra from niacin deficiency [109].

Interestingly, these two forms of malnutrition, which are not mutually exclusive, are frequently interconnected. A state of overweight/obesity is a known risk factor for the development and maintenance of micronutrient deficiencies [110], and the persistence of micronutrient deficiencies over time can trigger biochemical-metabolic alterations, promoting the development and subsequent maintenance of excessive weight gain [105].

Nutrition Interventions for Autistic Individuals

Many intervention strategies exist for autistic individuals, including the involvement of family members in meals, the consumption of a healthy diet and positive eating behaviors, and game-based exploration of visual, olfactory, and tactile food characteristics to expand the food repertoire [111].

Specifically, three main types of interventions for promoting a healthy eating pattern and addressing FS are described in the literature: (i) behavioral interventions, (ii) sensory-based interventions, and (iii) food chaining [112].

Despite the promising evidence highlighted by the scientific literature, no study has analyzed the long-term impact of interventions aimed at enhancing dietary quality and food choices in this vulnerable population through nutrition education. They have focused mainly on assessing the feasibility of improving symptoms related to social and behavioral areas, as well as dietary habits. In view of the high rate of FS observed in this vulnerable population, these programs should also incorporate a strategy for the management of FS to optimize its efficacy [112].

With reference to programs designed to reduce FS and expand the food repertoire, they can be classified into interventions for hyposensitivity and hypersensitivity (Fig. 2.6). There is currently no gold-standard approach, though behavioral strategies are associated with significant improvements in eating behavior. One limitation of these programs is the inconsistent introduction of healthy foods at the beginning of treatment sessions. Consequently, the path to improvement may be protracted, necessitating the later implementation of a program specifically targeting the enhancement of food quality [112]. For all these programs, a

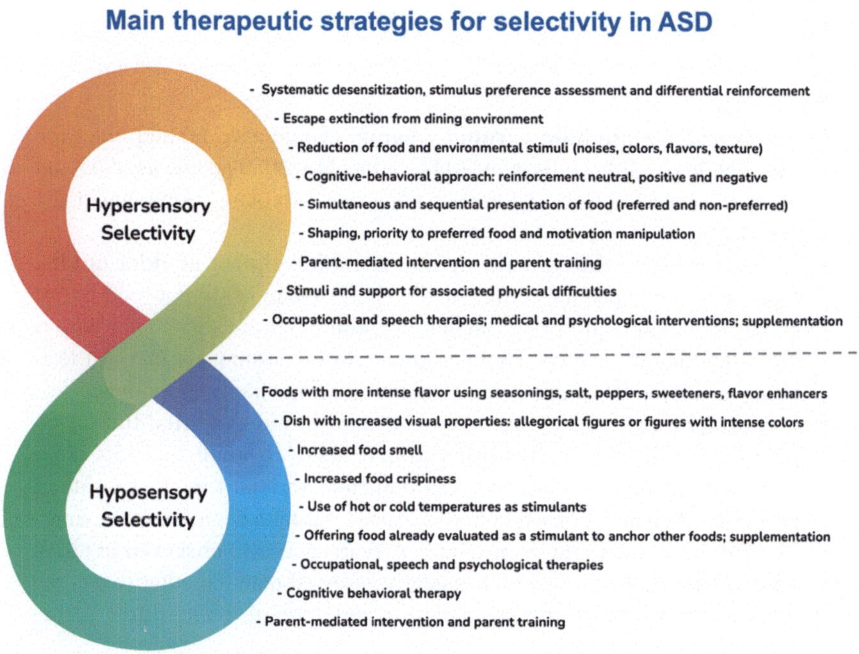

Fig. 2.6 Main intervention strategies for addressing FS in autistic individuals, depending on hypersensory or hypersensory perception. (Source: [111])

multidisciplinary approach is strongly suggested as a best practice to increase the effectiveness of interventions [111, 112]. Professionals involved in the interdisciplinary management of FS in autistic individuals are occupational and speech therapists, dietitians/nutritionists, psychologists, and medical doctors [92, 112].

Notably, parents and caregivers play a central role in these programs. Indeed, providing parental education and behavioral support, expanding family food preferences, and eliciting parental concerns are key aspects to be included in interventions for this target population [92]. Specifically, researchers have suggested that first there should be a focus on building healthy habits for the entire family before addressing FS [92].

Nonetheless, much remains to be done concerning nutritional intervention in autistic individuals and FS. To fill the existing research gaps, multifaceted, life-course interventions addressing outcomes of interest to autistic people and their families are needed [92].

In conclusion, considering the evidence presented, it is imperative to be aware of the specific needs of this target population to reduce their malnutrition risk by increasing their food repertoire. The first worldwide dietary recommendations have recently been formulated, providing a comprehensive discussion of the issues for all food groups and proposing feasible solutions for clinicians and companies [93]. To provide a concrete case study of FS in autistic individuals, the FOOD-AUT case study is analyzed in Chap. 4 (Section "Case Study: Pellegrini S.p.a. and FOOD-AUT Project").

Target IV: Malnutrition in Elderly People

Key Points

- **The world's population is rapidly aging**: people aged 60 and older are expected to reach 1.4 billion by 2030. In Italy, by 2050 people aged 65 and older are expected to be 3 times as numerous as young people under the age of 15;
- **Undernutrition-related malnutrition is not infrequent in older adults,** especially in the hospital setting, and it is due to a reduction of food intake and/or to inflammation (e.g., acute or chronic disease associated with a pro-inflammatory status). There are numerous determinants that can lead to the development of this form of malnutrition in elderly individuals;
- **Overweight and obesity are very common in older adults**. Indeed, in Italy 41% of the older adults suffer from being overweight, and 15% from obesity. Individual, social, and environmental variables interact to affect energy balance and, consequently, obesity, creating a complex framework.
- **Malnutrition due to micronutrients deficiency is also observed in older adults**. Nonetheless, it is less frequently assessed than the other two types of malnutrition. Thus, the prevalence of this type of malnutrition is currently underestimated.

The world's population is rapidly aging, with the number of people aged 60 and older expected to reach 1.4 billion by 2030 [113, 114]. Notably, by the mid-2030s, the number of older adults aged ≥80 years will outnumber the number of infants, whereas by the late 2070s, the number of people aged ≥65 years will surpass the number of children under 18 years [115].

The Italian scenario matches the global one: indeed, at the beginning of 2024 in Italy, the number of elderly individuals aged 65 and over was 14 million 358 thousand (approximately 25% of the total population), approximately twice as many children and young people under the age of 15 [57]. This number is predicted to further increase since, by 2050, people aged 65 and older are expected to be three times as numerous as young people under the age of 15 [57].

Thus, the increase in longevity is a tangible achievement of our time, representing one of the greatest challenges for the immediate future [116]. Only a systematic, urgent and adequate commitment at the socioeconomic, health and research levels will be able to turn this threat into an opportunity [117]. Indeed, there is an urgent need to promote healthy aging in this population to ensure the sustainability of the National Health System. First, it is crucial to define who the elderly are. There are two international definitions: the United Nations defines older adults as people aged 60 or 65 years or more, whereas the WHO defines people aged ≥65 years. However, the definition provided by the WHO is most used [118].

Focus: Changes Occurring with Aging

After the elderly population is defined, it is important to be aware of the changes that occur with age and how they affect nutritional status. Aging populations experience alterations in the gastro-intestinal tract, body composition changes, limitations in mobility, and low-grade inflammation [119, 120]. In the mouth, chewing disorders are frequently observed in elderly, frequently due to progressive tooth loss and the use of dentures. Reduced taste, smell and thirst perceptions are also reported, predisposing this population to a loss of appetite and reduced food intake, as well as to the use of excessive amounts of salt and sugar to give flavor to dishes. The aging process leads to delayed gastric emptying, reduced intestinal absorption of nutrients, and slower intestinal peristalsis. Indeed, older adults are more prone to develop nutritional deficiencies and to suffer from constipation. Furthermore, the gut microbiota significantly changes during aging, with a general decrease in microbial diversity, leading to the development of dysbiosis, which contributes to multiple pathological conditions associated with aging [121]. Through the aging process, gradual physiological changes in body composition occur. There is a progressive decrease in fat-free mass (including muscle mass) and a concomitant increase in fat mass (including visceral fat mass), which is also visible from the increased waist circumference. Indeed, muscle mass and strength (in parallel with bone mineral density) peak in young adulthood and, after a plateau, begin to decline gradually, with a faster decline in strength [122]. Furthermore, the increase in body fat mass, especially at the visceral level, significant impacts on the risk of developing cardiovascular diseases [123].

Owing to the impact of aging on muscle and bone mineral density, elderly have a greater risk of falls and fractures than adults do [124]. They also experience a reduced balance capacity, worsening this scenario. Thus, older adults are more prone to experience limitations in mobility and activities of daily living, a loss of independence, and the development of disability [125]. Finally, an age-related increase in markers of inflammation has been reported in aging populations, which could underlie the reduced ability of elderly to cope with various stressors, increasing the risk of disease, disability and death. Therefore, the term "inflammaging" was created in 2000 by Prof. Franceschi of the University of Bologna [126] and has been endorsed by the research community [119, 127].

Considering all the above, it is not surprising that older people are listed among vulnerable populations, which are groups of individuals at increased risk for health problems and health disparities [128]. Specifically, compared with adults, older adults have a greater risk of developing malnutrition [129, 130].

Nutritional Status in Older Adults

There are three forms of malnutrition that may affect, even simultaneously, an older person: undernutrition, overnutrition (e.g., overweight or obesity), and micronutrient deficiency (also known as "hidden hunger") [4, 5, 131].

Undernutrition was defined in 2019 by the Global Leadership Initiative on Malnutrition (GLIM) as "a state resulting from lack of nutrient intake or absorption that leads to altered body composition and body cell mass, resulting in decreased physical and mental function and impaired clinical outcome of a disease" [132]. It is diagnosed by the clinician when at least one phenotypic criterion is fulfilled (i.e., low BMI, rapid involuntary weight loss or reduced muscle mass), combined with at least one etiologic criterion (i.e., reduced food intake or assimilation or disease burden/inflammatory conditions).

Delving into the assessment of phenotypic criteria, weight and height can be measured through a calibrated weighing scale and a stadiometer, respectively. BMI is subsequently calculated as weight/(height)^2, and it is considered low in Europeans when it is <20 kg/m^2 in people <70 years old or < 22 kg/m^2 in people ≥70 years old. The weight loss (%) is calculated as the difference between the weight of the previous 6 months and the current weight, divided by the initial weight. It satisfies the criterion when ≥5%. Notably, the same rate of weight loss does not fulfill the GLIM criterion when it is due to a diet, since in this case, it is voluntary. Muscle mass quantity is evaluated through specific instruments (e.g., dual energy X-ray absorptiometry (DXA) and bioelectrical impedance analysis (BIA)), which are not available for the general population. BIA is an instrument that generates a small electrical current that passes through the body, and the body's resistance is measured [133, 134]. The values (resistance and reactance) provided by BIA are then inserted into an equation specific for Caucasians to estimate the amount of fat-free mass [135]. The value obtained is compared to the sex-specific cutoff provided by the GLIM consensus [136].

With respect to the etiologic criteria, reduced food intake or assimilation can be assessed through a structured interview, combined with the evaluation of gastrointestinal symptoms (e.g., dysphagia, nausea, vomiting, diarrhea, and constipation) and the disease burden of the older adult. Indeed, some diseases (e.g., short bowel syndrome, pancreatic insufficiency) and surgical interventions (e.g., bariatric surgery) could lead to malabsorption. This criterion is satisfied when the energy intake is ≤50% of the energy requirement of the person for more than 1 week, when there is any reduction for more than 2 weeks, or when any chronic gastrointestinal condition that adversely impacts food assimilation or absorption. The second possible etiologic criterion is the presence of acute disease/injury or chronic disease-related disease. Severe inflammation or moderate inflammation are likely associated with many acute or chronic diseases (e.g., major infection, burns, trauma, cancer, chronic obstructive pulmonary disease, or congestive heart failure). When available, C-reactive protein may be used as a supportive laboratory measure [137].

Many factors contribute to the development of malnutrition in older adults, as highlighted by the model "Determinants of Malnutrition in Aged Persons" (DoMAP) created within the European Knowledge Hub Malnutrition in the Elderly (MaNuEL) [129, 130].

Unfortunately, the prevalence of malnutrition in older adults is frequently underreported. Most recent European data show a prevalence ranging from 0.8% (Netherlands and Sweden) to 11.0% (Italy and France) [138]. In Italy, the SIMI-NUTRO study has been recently conducted [139]. It was a multicenter, observational cohort study conducted in 16 Internal Medicine Wards located in North, Center and South of Italy. The aim was to assess malnutrition risk and diagnosis by MUST (Malnutrition Universal Screening Tool), MNA-SF (Mini Nutritional Assessment Short-Form), and GLIM, respectively. A high prevalence of malnutrition risk (42.3% and 26.3%, respectively) and diagnosis (37.3%) has been observed [139]. Besides this study, more European and Italian data on the prevalence of malnutrition according to the GLIM criteria in other settings, e.g., community and nursing homes, is urgently needed [140, 141].

Shifting to overnutrition (i.e., overweight and obesity), they have been defined by the WHO as "abnormal or excessive fat accumulation that presents a risk to health" [113, 114]. Obesity is a complex, heterogeneous, chronic, and progressive disease that substantially affects health, quality of life, and mortality [142].

With respect to the WHO diagnosis, a person with a BMI between 25 and 29.9 kg/m^2 suffers from being overweight (sometimes referred to as "preobese"), whereas a person with a BMI $\geq$ 30 kg/m^2 is considered with obesity. Specifically, a BMI between 30.0–34.9 kg/m^2 corresponds to class I obesity, a BMI between 35.0–39.9 kg/m^2 corresponds to class II obesity, and a BMI $\geq$ 40 kg/m2, corresponds to class III obesity [143, 144]. Thus, overnutrition-related malnutrition can be diagnosed by measuring weight and height and by calculating weight/(height)^2, as previously explained [14, 15].

Other anthropometric measurements can be used to assess excessive abdominal adiposity, such as waist circumference. It can be collected through an inelastic tape measure at the midpoint between the lower margin of the last rib and the upper margin of the iliac crest [145]. However, given the difficulty in finding this reference point in people with obesity, waist circumference is measured at the level of the belly button. The value obtained is compared with sex- and ethnicity-specific cut-offs promoted by the International Diabetes Federation [146]. The values specific for European people indicating excessive abdominal adiposity are $\geq$94 cm in men and $\geq$80 cm in women. Measures above these two sex-specific values denote a greater cardiovascular risk. Furthermore, it is possible to calculate the waist-to-height ratio (WHtR = waist/height). Higher values of WHtR ($\geq$0.5) indicate a greater risk of obesity-related cardiovascular diseases [147], which are correlated with central obesity.

Many aspects participate in the development of overweight or obesity in aging populations, creating a complex framework. These factors influencing obesity are graphically represented in the Foresight obesity system map of the United Kingdom

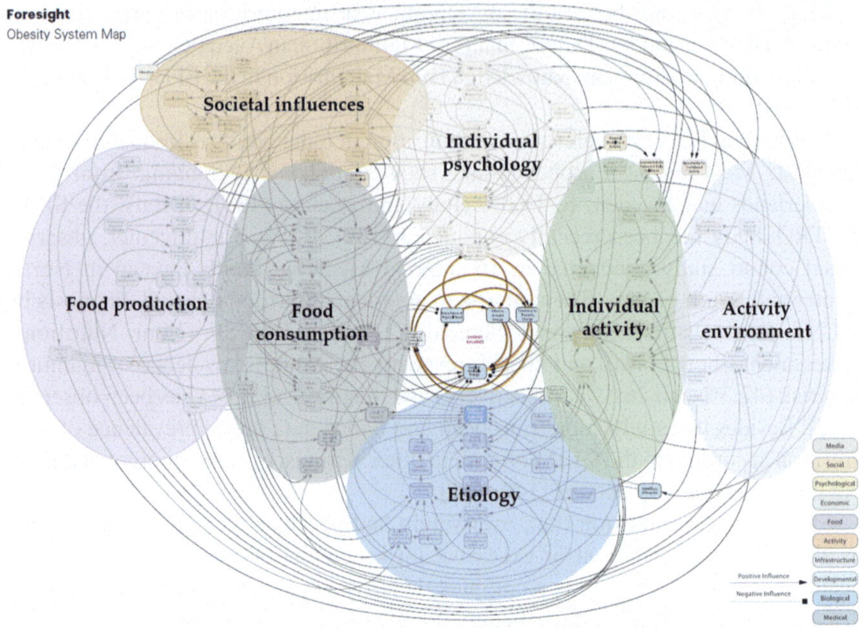

Fig. 2.7 U.K. Foresight obesity system map highlighting the complex existing network of factors influencing obesity. (Source: [148])

(Fig. 2.7), which aims to illustrate how various individual, social, and environmental variables interact to affect energy balance and, consequently, obesity.

Obesity has dramatically increased in prevalence worldwide, as clearly highlighted by the data presented in the Lancet journal, referring to the period 1990–2022 [17]. Owing to the magnitude of the problem, the WHO has declared it a global epidemic [142] and has launched the WHO acceleration plan to stop obesity [9, 10]. In continental Europe, the prevalence of obesity in older adults is high, 33.6% [149]. Italian data provided by the National Surveillance System Passi d'Argento related to the 2022–2023 biennium show that, in Italy, most people over 65 years of age (56%) were overnourished: 41% were overweight, and 15% were living with obesity [150].

The third form of malnutrition is malnutrition due to micronutrient deficiency, which is defined by the ESPEN guidelines as "the evidence of objective loss of a micronutrient in body fluids, or intake below standard recommendation" and "the presence of clinical signs or symptoms, compatible with a micronutrient deficiency or blood/plasma concentrations below reference range together with metabolic effects of inadequacy" [151]. Notably, it frequently coexists with the other two forms of malnutrition [131].

For the general population, international recommendations are available, i.e., recommended dietary allowances or dietary reference intakes. Nonetheless, to date, no procedures for the determination of requirements or recommendations are available for patients with acute and chronic diseases, although micronutrient deficiencies are frequently observed in this population [151].

Micronutrient deficiencies are assessed by blood analysis combined with the investigation of their dietary intake through standardized interviews conducted by nutritional professionals as well as specific signs and symptoms [120, 151].

With respect to prevalence data in elderly individuals, a systematic review [152] analyzed the most common micronutrient deficiencies in community-dwelling older adults. Among the 20 nutrients analyzed, vitamin D, thiamin (vitamin B1), riboflavin (vitamin B2), calcium, magnesium and selenium were considered possible public health concerns in this target population. Notably, micronutrient deficiencies are even more neglected than undernutrition-related malnutrition. This is probably because blood exams are costly and because there is a need for qualified nutrition professionals to understand their intake [152, 153]. As far as the Italian scenario is concerned, there is a lack of updated data on micronutrient deficiencies in older adults. One of the most recent Italian publications on this topic is the Mediterranean Healthy Eating, Aging, and Lifestyle (MEAL) study [154].

In summary, the prevalence rates of malnutrition in its three forms in older adults (especially undernutrition-related and overnutrition-related malnutrition) clearly deserve public health attention. Considering the rapid aging of the global population, there is indeed a need to address malnutrition in this vulnerable population [124, 155]. Detailed dietary recommendations on how to treat malnutrition in elderly are beyond the scope of this monograph. In any case, it is also important for the general population to have basic knowledge of these issues to suspect the presence of one or more of these forms of malnutrition in elderly individuals. The next step involves referral to a qualified nutrition professional, who will proceed with the eventual diagnosis and develop appropriate treatment.

In light of this scenario, the ongoing cross-sectional phase of the SENIOR study (Sustainable Eating patterN to lImit malnutrition in oldeR adults) is briefly described as a virtuous case study for assessing the TBM in older adults within the hospital setting, enrolling both inpatients and outpatients (Box 2).

Box 2 Brief Description of the Cross-Sectional Phase of the SENIOR Project, Conducted by the Laboratory of Dietetics and Clinical Nutrition (LDNC) of the University of Pavia, in Collaboration with ICS Maugeri IRCCS Hospital, Fondazione IRCCS Policlinico san Matteo and Fondazione IRCCS San Gerardo Dei Tintori di Monza

The **SENIOR study ("Sustainable dietary patterns to counter malnutrition in the elderly")** is an ongoing multicenter study funded by National Recovery and Resilience Plan [156]. Its first cross-sectional phase aims to i) investigate the prevalence of malnutrition and sarcopenia in hospitalized (inpatients) and community-dwelling (outpatients) older adults, and ii) better understand the risk factors that are associated with the development of these two conditions. Specifically, malnutrition is assessed in its three forms: undernutrition according to the GLIM criteria (using the MUST as screening tool), overweight and obesity according to the WHO criteria, and micronutrient deficiencies (i.e., vitamin D, B9, B12, Na, K, Ca, Mg, Fe, Zn) based on blood exams. Furthermore, data are also collected on lifestyle (adherence to healthy dietary patterns, physical activity, sleep quality, smoking and alcohol consumption), the social sphere (e.g. social support), quality of life, independence, depression and blood biomarkers, including cytokines. In mid-September 2025, 274 participants were enrolled in the cross-sectional phase (228 inpatients and 46 outpatients) at the ICS Maugeri IRCCS in Pavia, Italy. Results will be published in 2026.

Target V: Malnutrition in Hospitalized Patients

Key Points

- **The skeleton is still in the hospital closet:** despite the efforts in the last decades, the prevalence of malnutrition in the hospitals is still very high (39% with the GLIM criteria). Furthermore, the prevalence of malnutrition in the hospital varies according to the ward and the pathology the patient suffers from;
- **Currently the hospital is not a "nutrition-friendly" setting.** Indeed, numerous factors relating to the individual, to hospital routines and therapies, to inadequate training of healthcare professionals and to hospital food service play a role in malnutrition development, maintenance and worsening;
- **The evidence on the association between hospital malnutrition and adverse clinical outcomes** (e.g., higher risk of mortality, higher length of stay and more frequent hospital readmission) **is strong.** Thus, this justifies urgent public health initiatives.
- **Therefore, every patient should be screened for malnutrition at hospital admission**, independently of their medical condition. **A positive screening should be always followed by a complete diagnostic assessment and a personalized dietary intervention to tackle malnutrition.**

With advances in medical treatment and the increasing number of patients who are older than 65 years and have multiple morbidities, disease-related malnutrition in patients with multiple illnesses has become a growing concern [157].

The first landmark publication on malnutrition in hospitalized patients was published in 1974 by Butterworth Jr. CE with the evocative title "*The skeleton in the hospital closet*" [158], highlighting high rates of nutritional deficiencies in both medical and surgical patients. These articles raised researchers' attention to the topic, and many articles were published thereafter [159, 160]. Despite these effort, malnutrition remains a central healthcare problem and a public health concern.

Prevalence of Malnutrition in the Hospital Setting

Malnutrition is common in the hospital setting, particularly among older adults [161, 162]. Indeed, up to one-third of patients in hospitals are malnourished or at risk of malnutrition at the time of hospital admission [157]. Furthermore, the reported prevalence of malnutrition diagnosed by the GLIM criteria in hospitalized adults is 39% [141]. Recent Italian data provided by the SIMI-NUTRO study showed a high prevalence of malnutrition risk (42.3% according to the MUST and 26.3% to the MNA-SF, respectively) and diagnosis (37.3% according to GLIM criteria) in 16 Internal medicine wards [139]. However, there is still a lack of national prevalence data of malnutrition according to the GLIM criteria in other hospital wards and outpatient clinics. In fact, other recent National data, published by the Società Italiana di Nutrizione Artificiale e del Metabolismo (SINPE), do not refer to the GLIM criteria for diagnosis [155]. The reported prevalence was approximately 50% of the hospitalized adult population.

In medical wards, the prevalence of malnutrition according to the GLIM criteria is greater in surgical wards (47%) than in internal medicine wards (40%) [141]. Wards differences have also been highlighted in the Italian context: 36.8–63.4% in medicine, 29.3–70.2% in surgery, and 47.5–51.8% in oncology [155]. Notably, the GLIM criteria have undergone specific validation for intensive care units [163] and for emergency wards [164].

With respect to disease type, the reported prevalence in cancer patients is 36–52%, whereas that in acute and critically ill patients is 33–56% [141]. Indeed, malnutrition rates are reportedly high in patients with specific diseases [132], such as cancer [165], inflammatory bowel disease [166], chronic obstructive pulmonary disease [167], abdominal surgery [168], diabetes [169], chronic kidney disease [170], neurological diseases and dysphagia [171, 172]. Of note, the GLIM criteria have also been validated for critically ill patients [173, 174] and those with cancer [175].

How Can Hospitals Contribute to the Development of Malnutrition in Patients?

The hospital setting plays a central role in malnutrition development (Fig. 2.8), since there is moderate evidence that hospitalization is a determinant of malnutrition [176]. Two main mechanisms have been defined: inflammation-driven catabolism and undernutrition-driven catabolism [157]. Indeed, a patient's nutritional status often deteriorates during the hospital stay because of illness-related loss of appetite, drug-related side effects, and fasting orders for diagnostic studies. Furthermore, diseases that impair the normal functioning of the digestive system and disease-related wasting are known to contribute to worsening malnutrition [124, 157].

In terms of the role of disease burden in malnutrition development, acute and chronic inflammation are known key contributing factors [132]. Proinflammatory cytokines (e.g., interleukin-6 and tumor necrosis factor α) affect brain circuits that control food intake, favoring a reduced appetite with a decrease in the intake of energy and protein. They also act on muscle, favoring catabolic processes and muscle loss, leading to patient fatigue and poor energy and strength [124, 157]. Thus, inpatients often develop difficulties in walking (e.g., reduced balance capacity and strength) and performing daily activities, resulting in the loss of independence and immobilization [124, 161, 162].

Notably, hospitals are also renowned for contributing to further declines in nutritional status during hospitalization. Diseases (e.g., acute illness and injuries), inadequate meal service (e.g., inflexible mealtimes, limited food choice, insufficient time to eat meals), and insufficient amounts of food provided compared with patients' higher energy and protein requirements are all factors that can exacerbate weight loss during hospital stays [124, 157].

Considering all the above, many factors contribute to making the hospital a favorable setting for developing and worsening malnutrition. These barriers are due

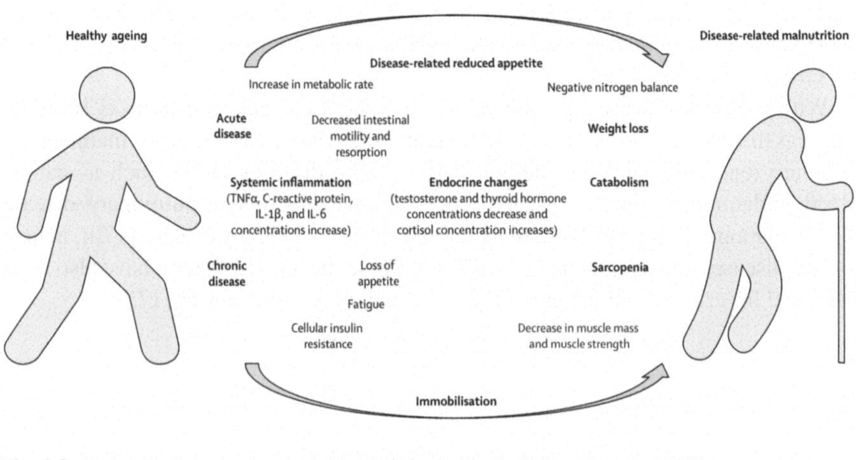

Fig. 2.8 Pathophysiology of malnutrition in hospitalized adults. (Source: [157])

to hospital routines (e.g., mealtime interruptions and procedure-related fasting) and therapies, inadequate training of healthcare professionals (e.g., inadequate malnutrition screening, lack of a figure performing it, poor importance attributed to nutrition-related issues), and hospital food services (e.g., meal dissatisfaction, meals not adequate for patients' chewing difficulties) [157, 161, 162, 177, 178]. Other barriers are related to the patient's health history throughout his/her hospital stay, such as the effect of illness, the development of chewing difficulties (e.g., due to dysphagia), poor appetite, clinical impairment up to dementia, and low mood/ depression [161, 162, 172, 176, 177]. Nonetheless, few projects have attempted to make the hospital a more "nutrition-friendly" setting, e.g., providing protected mealtimes and feeding assistance [179–181].

The Impact of Malnutrition on the Health Status of Hospitalized Patients

The evidence on the association between hospital malnutrition and adverse clinical outcomes is strong [157], justifying urgent public health initiatives. These include increased morbidity and mortality, functional decline, falls and fractures [124, 157]. The mortality risk within 1 year for hospitalized malnourished patients diagnosed with the GLIM criteria is 2.62 times greater than that for nonmalnourished patients [141]. Moreover, the risk of death beyond 1 year in malnourished patients diagnosed by the GLIM criteria remains double that in patients with a normal nutritional status [141].

Furthermore, longer lengths of stay (LOSs) and hospital complications and higher rates of hospital readmission are reported in malnourished hospitalized patients [124, 157]. This clearly highlights the importance of tackling malnutrition as soon as possible to prevent a malnutrition-hospitalization cycle (Fig. 2.9), which is detrimental to the patient's health [182].

Notably, the associations between malnutrition and adverse clinical outcomes are largely independent of the underlying medical condition [157]. This emphasizes the role played by the hospital setting.

Which Are the Proper Health Care Practices for Tackling Malnutrition in the Hospital Setting?

There are two main steps [132, 183, 184]:

1. Attention should be given to diagnosing malnutrition at hospital admission and as soon as possible thereafter. Indeed, a patient can be well nourished at hospital admission and develop malnutrition within weeks. Validated screening tools for adults, e.g., the Mini Nutritional Assessment–Short Form (MNA-SF) and the

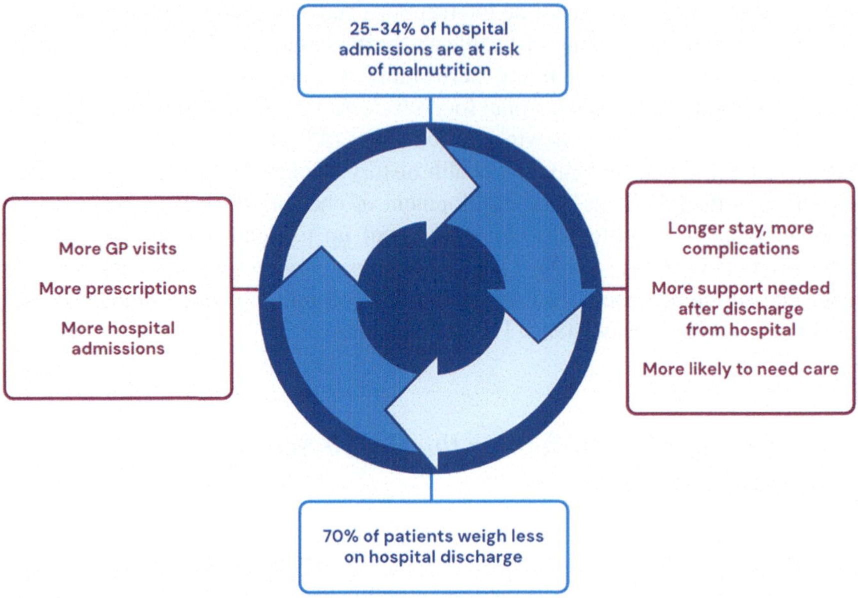

Fig. 2.9 Malnutrition–hospitalization cycle. (Source: [182])

Malnutrition Universal Screening Tool (MUST) (Box 3), as well as the Nutritional Risk Screening 2002 (NRS-2002), exist in Italy [185]. They should be routinely used in the hospital setting since they are also part of the diagnostic process according to the GLIM criteria. These screening tools are useful for the rapid, early identification of malnutrition risk. Every patient should be screened independently of their medical condition [186]. When a risk is observed (i.e., positive result of the screening tool), malnutrition assessment should be performed on the basis of the GLIM criteria (see Section "Target IV: Malnutrition in Elderly People");

2. After malnutrition diagnosis, the patient should receive a personalized nutritional intervention performed by a nutrition professional. The nutrition care plan should be adapted according to the patient's needs, e.g., energy, macronutrient and micronutrient requirements (also depending on the specific disease), food preferences and chewing capacity. The definition of a proper nutritional intervention for malnourished patients is beyond the scope of this chapter but should be based on internationally recognized guidelines [187].

However, are these two steps worth the effort, since they require investment of time and money? A recent umbrella review on the effectiveness of nutritional intervention in hospitalized patients with or at risk of malnutrition revealed that nutritional intervention is effective for mortality outcomes at 30 days, 6 months, and 12 months [188]. Furthermore, other systematic reviews highlighted the benefits of nutrition interventions in hospitalized elderly individuals [189–192].

The most important study that has been conducted on hospitalized patients is the EFFORT trial (Effect of early nutritional support on Frailty, Functional Outcomes, and Recovery of malnourished medical inpatients Trial) [186]. It included 2088 polymorbid inpatients at nutritional risk in 8 Swiss hospitals. As a result, the use of individualized nutritional support during the hospital stay improved important clinical outcomes (e.g., survival) compared with standard hospital food. Therefore, the evidence provided by this study supported the concept of systematically screening medical inpatients at hospital admission regarding nutritional risk, followed by a nutritional assessment and a personalized nutritional intervention to address malnutrition [186]. Furthermore, many ancillary publications have been produced by the EFFORT research group, highlighting the effects on patients with different diseases, e.g., chronic heart failure [193], chronic kidney disease [194, 195], diabetes [196] and cancer [194, 195].

In conclusion, given the risk factors for malnutrition in hospitals and the high reported prevalence, the importance of screening, diagnosing and treating malnutrition as early as possible clearly emerges. The Lombardy Region represents a virtuous case study for brief analysis (Box 4).

Box 3 Malnutrition Screening Tools for Hospitalized Patients in Italy

The Mini Nutritional Assessment—Short Form (MNA-SF) includes six questions from the full version of the MNA [197]. The total score ranges from 0 to 14 points. A total score between 12–14 indicates a normal nutritional status, between 8–11 a risk of malnutrition and <7 a high risk of malnutrition [198, 199]. The Malnutrition Universal Screening Tool (MUST) was developed by the British Association for Parenteral and Enteral Nutrition (BAPEN) as a five-step screening tool. The total score ranges from 0 to 6 points. A total score of 0 indicates a low malnutrition risk, and of 1 a moderate risk, while a score ≥2 suggests a high malnutrition risk [200]. The full texts of the questionnaires are available at the following references: [201, 202].

Box 4 Best Practices Implemented in Italy for Systematically Diagnosing and Treating Malnutrition in Hospitalized Patients: The Lombardy Case Study

In July 2020, the **Italian Ministry of Health** highlighted the lack of a nationwide system for screening and diagnosis of malnutrition in hospitals, proposing operational solutions [203]. With **DGR XII/1812 of 29/01/2024**, the Region of **Lombardy** became the first Region to activate mandatory nutritional screening pathways in the facilities of the regional health system (including hospitals) and at home [204] through the use of validated questionnaires (e.g. MUST). In addition, according to this resolution, each positive screening should be followed by the definition of a risk-adapted care pathway. By **31 December 2024**, this nutritional screening has been systematically activated for patients admitted to ordinary hospitals in Lombardy. In **June 2025** Regional guidelines for conducting nutrition screening and managing the risk of malnutrition were released [205]. Noteworthy, the **Clinical Nutrition Network of Lombardy** has been established to help overcome existing barriers towards effective implementation [206].

Target VI: Malnutrition in the Workplace: Organizations' Employees

Workplace health promotion is defined as "the combined efforts of employers, employees, and society to improve the health and well-being of people at work" [207]. Since individuals spend the majority of their day at work—approximately 7.5 hours daily in Europe—the significance of promoting health in the workplace becomes clear. The WHO defines a healthy workplace as one where employees and management work together—guided by identified needs—in an ongoing process to safeguard and enhance the health, safety, and well-being of all workers while also supporting the organization's sustainability. Such a setting should actively address lifestyle-related risks, including sedentary behavior, psychosocial stress, and poor nutrition, while also discouraging smoking and alcohol use [143, 144]. Moreover, a truly healthy workplace should promote the adoption of healthy lifestyle habits within employees' families, thereby positively influencing the broader community's health. Nevertheless, workplace-based interventions aimed at improving fitness and nutrition—particularly those incorporating work-related outcomes—are inherently complex, making their implementation and accurate assessment of their effectiveness difficult. Workplace intervention studies have generally focused on goals such as weight loss, smoking cessation, reducing alcohol consumption, and promoting healthier eating habits or increased physical activity. Several of these studies reported measurable improvements in diet and body weight [208, 209]. Although these outcomes were seen as encouraging, the 2015–2020 Dietary Guidelines for Americans noted that the effects were inconsistent and modest in scale. Thus, a better understanding of the issues concerning malnutrition and affecting employees represents a key step for organizations to learn how to address them more effectively.

Among the types of malnutrition, overweight is the one that has been covered the most in the literature on health in organizations. Key workplace-related risk factors contributing to obesity include high job stress, long working hours, night shifts, poor nutritional offerings (e.g., unhealthy cafeteria or vending options), sedentary routines, and environments that discourage physical activity [210]. Importantly, many of these factors are modifiable and serve as risk factors for NCDs [211, 212]. Consequently, most WHP programs aim to address key areas such as healthy eating, weight control, physical activity, smoking cessation, alcohol reduction, and sleep improvement, aligning with the WHO and European health strategies [211, 212].

Notably, some workers, such as shift workers and truck drivers, are at increased risk of developing overnutrition-related malnutrition. Indeed, a higher prevalence of overweight and obesity is reported in employees with a day/night rotational work schedule, as well as a higher frequency of meal intake or poor nutritional quality/habits, than in day workers [213]. For instance, truck driving is a job associated with a high risk of developing overweight and obesity due to irregular shift work, prolonged driving, sleep deprivation, and limited access to healthy food. These aspects favor the adoption of unhealthy lifestyles, including poor dietary choices, sedentary behavior, smoking and excessive alcohol, which are commonly observed in this worker population [214].

Instead, undernutrition-related malnutrition is poorly investigated in the literature on health and organizations. Studies investigating employment patterns among individuals with eating disorders (EDs) reveal a negative effect on work participation and performance, ranging from reduced productivity to increased risk of unemployment. The focus of most studies lies mainly on underweight due to eating disorders (e.g., anorexia nervosa) and not malnutrition according to the GLIM criteria or other criteria. The systematic review by Safi et al. [215] highlights some empirical evidence on the relationship between EDs and work participation. However, it is often suggested that EDs impair workplace functioning due to low body weight, which can lead to nutritional deficiencies and reduced cognitive and physical capacity. However, low weight alone is unlikely to fully explain the impact of EDs on work. Difficulties in work participation and performance are also reported in individuals considered recovered—those with normalized weight and eating behaviors—as well as in those with EDs not marked by low weight, such as bulimia nervosa. Furthermore, similar impairments are observed in other mental health conditions, including depression, bipolar disorder, and anxiety, suggesting that broader psychological factors may be at play. Additionally, the authors' meta-analysis revealed that diet has a limited effect on work performance. However, given that EDs involve both physical and psychological impairments, their impact on workplace functioning may differ from that of other mental health conditions. A deeper understanding of this relationship is key to recognizing how EDs influence daily functioning.

Finally, the impact of micronutrient deficiency on workers has been poorly investigated. Previous studies have focused mostly on vitamin D deficiency [216]. A sedentary lifestyle and the indoor workplace have emerged as key barriers to sun exposure, thus preventing individuals from reaching an optimal level of vitamin D [217]. More studies on malnutrition from a health perspective in organizations are needed to investigate the aspects of undernutrition and nutrient deficiency affecting employees in organizations and consequently implementing proper practices to address them.

References

1. Murray CJ, Aravkin AY, Zheng P, Abbafati C, Abbas KM, Abbasi-Kangevari M et al (2020) Global burden of 87 risk factors in 204 countries and territories, 1990–2019: a systematic analysis for the Global Burden of Disease Study 2019. Lancet 396(10258):1223–1249
2. Mannar V, Micha R, Allemandi L, Afshin A, Baker P, Battersby J, … Schofield D (2020) 2020 global nutrition report: action on equity to end malnutrition. 89023
3. Shrimpton R, Rokx C (2012) The double burden of malnutrition. In: A review of global evidence. World Bank, Washington, DC
4. World Health Organization (2024) Malnutrition fact sheets. Available at: https://www.who.int/health-topics/malnutrition#tab=tab_1. Accessed 01 Feb 2025
5. World Health Organization (2024) Malnutrition. World Health Organization. Available at: https://www.who.int/news-room/fact-sheets/detail/malnutrition. Accessed 29 Apr 2025 [online]

6. World Health Organization (2006) Weight-for-length/height child growth standards. Available at: https://www.who.int/tools/child-growth-standards/standards/weight-for-length-height. Accessed 1 Feb 2025
7. FAO, IFAD, UNICEF, WFP, WHO (2024) The state of food security and nutrition in the world 2024—financing to end hunger, food insecurity and malnutrition in all its forms, Rome
8. Han X, Ding S, Lu J, Li Y (2022) Global, regional, and national burdens of common micronutrient deficiencies from 1990 to 2019: a secondary trend analysis based on the Global Burden of Disease 2019 study. EClinicalMedicine 44
9. World Health Organization (2023) Accelerating aneaemia reduction: a comprehensive framework for action. World Health Organization
10. World Health Organization (2023) WHO acceleration plan to stop obesity. [online] WHO. Available at: https://iris.who.int/handle/10665/370281. Accessed 29 Apr 2025
11. Stevens GA, Bennett JE, Hennocq Q et al (2015) Trends and mortality effects of vitamin A deficiency in children in 138 low-income and middle-income countries between 1991 and 2013: a pooled analysis of population-based surveys. Lancet Glob Health. 3(9):e528–e536. https://doi.org/10.1016/S2214-109X(15)00039-X
12. Andersson M, Karumbunathan V, Zimmermann MB (2012) Global iodine status in 2011 and trends over the past decade. J Nutr 142(4):744–750
13. Rubino F, Batterham RL, Koch M, Mingrone G, le Roux CW, Farooqi IS et al (2023) Lancet Diabetes & Endocrinology Commission on the definition and diagnosis of clinical obesity. Lancet Diabetes Endocrinol 11(4):226–228
14. Busetto L, Dicker D, Frühbeck G, Halford JC, Sbraccia P, Yumuk V, Goossens GH (2024) A new framework for the diagnosis, staging and management of obesity in adults. Nat Med 30(9):2395–2399
15. Rubino F, Cummings DE, Eckel RH, Cohen RV, Wilding JP, Brown WA et al (2025) Definition and diagnostic criteria of clinical obesity. Lancet Diabetes Endocrinol 13(3):221–262
16. World Obesity Federation (2024) World obesity atlas 2024. World Obesity Federation, London
17. Phelps NH, Singleton RK, Zhou B, Heap RA, Mishra A, Bennett JE et al (2024) Worldwide trends in underweight and obesity from 1990 to 2022: a pooled analysis of 3663 population-representative studies with 222 million children, adolescents, and adults. Lancet 403(10431):1027–1050
18. Popkin BM, Corvalan C, Grummer-Strawn LM (2020) Dynamics of the double burden of malnutrition and the changing nutrition reality. Lancet 395(10217):65–74
19. Gómez MI, Barrett CB, Raney T, Pinstrup-Andersen P, Meerman J, Croppenstedt A et al (2013) Post-green revolution food systems and the triple burden of malnutrition. Food Policy 42:129–138
20. Popkin BM (2014) Nutrition, agriculture and the global food system in low- and middle-income countries. Food Policy 47:91–96
21. Seenivasan S, Talukdar D, Nagpal A (2023) National income and macro-economic correlates of the double burden of malnutrition: an ecological study of adult populations in 188 countries over 42 years. Lancet Planet Heath 7(6):e469–e477
22. Popkin BM (2017) Relationship between shifts in food system dynamics and acceleration of the global nutrition transition. Nutr Rev 75(2):73–82. https://doi.org/10.1093/nutrit/nuw064
23. De Boo HA, Harding JE (2006a) The developmental origins of adult disease (Barker) hypothesis. Aust N Zeal J Obstet Gynaecol 46(1):4–14
24. De Boo HA, Harding JE (2006b) The developmental origins of adult disease (barker) hypothesis. Aust N Zeal J Obstet Gynecol 46(1):4–14
25. Baker P, Machado P, Santos T, Sievert K, Backholer K, Hadjikakou M et al (2020) Ultra-processed foods and the nutrition transition: global, regional and national trends, food systems transformations and political economy drivers. Obes Rev 21(12):e13126
26. Forde CG (2023) Beyond ultra-processed: considering the future role of food processing in human health. Proc Nutr Soc 82(3):406–418

27. Imamura F, Micha R, Khatibzadeh S, Fahimi S, Shi P, Powles J, Mozaffarian D (2015) Dietary quality among men and women in 187 countries in 1990 and 2010: a systematic assessment. Lancet Glob Health 3(3):e132–e142
28. Monteiro CA, Cannon G, Levy RB, Moubarac JC, Louzada ML, Rauber F et al (2019) Ultra-processed foods: what they are and how to identify them. Public Health Nutr 22(5):936–941
29. Swinburn BA, Kraak VI, Allender S, Atkins VJ, Baker PI, Bogard JR et al (2019) The global syndemic of obesity, undernutrition, and climate change: the lancet commission report. Lancet 393(10173):791–846
30. CREA (2018) Linee Guida per una Sana Alimentazione
31. Mistura L, Le Donne C, D'Addezio L, Ferrari M, Comendador FJ, Piccinelli R, … IV SCAI field work team. (2025) The Italian IV SCAI dietary survey: Main results on food consumption. Nutrition, metabolism, and cardiovascular diseases: NMCD, 103863
32. World Obesity Federation, Italy (2026) Global Obesity Observatory. https://data.worldobesity.org/country/italy-102/. Accessed 01 Apr 2025
33. Global Nutrition Report, Italy (2026) Country nutrition profile. Available at: https://globalnutritionreport.org/resources/nutrition-profiles/europe/southern-europe/italy/. Accessed 1 Feb 2025
34. United Nations Children's Fund (UNICEF) (2024) Child food poverty. Nutrition Deprivation in Early Childhood. Child Nutrition Report. UNICEF, New York, June 2024
35. Siddiqui F, Salam RA, Lassi ZS, Das JK (2020) The intertwined relationship between malnutrition and poverty. Front Public Health 8:453
36. Francis L, DePriest K, Wilson M, Gross D (2018) Child poverty, toxic stress, and social determinants of health: screening and care coordination. Online J Issues Nurs 23(3):2
37. Tanumihardjo SA, Anderson C, Kaufer-Horwitz M, Bode L, Emenaker NJ, Haqq AM et al (2007) Poverty, obesity, and malnutrition: an international perspective recognizing the paradox. J Am Diet Assoc 107(11):1966–1972
38. Popkin BM, Reardon T (2018) Obesity and the food system transformation in Latin America. Obes Rev 19(8):1028–1064
39. Jaacks LM, Vandevijvere S, Pan A, McGowan CJ, Wallace C, Imamura F et al (2019) The obesity transition: stages of the global epidemic. Lancet Diabetes Endocrinol 7(3):231–240
40. Kagenaar E, Van Hemelrijck WMJ, Kunst AE, Janssen F (2022) Long-term trends in obesity prevalence by socioeconomic group in five European countries and the USA: the relevance of the diffusion of innovations theory. Obes Facts 15(6):753–761
41. Robertson A, Loring B (2014) Obesity and inequities. Guidance for addressing inequities in overweight and obesity. World Health Organization
42. Monteiro CA, Conde WL, Lu B, Popkin BM (2004) Obesity and inequities in health in the developing world. Int J Obes 28(9):1181–1186
43. Reyes Matos U, Mesenburg MA, Victora CG (2020) Socioeconomic inequalities in the prevalence of underweight, overweight, and obesity among women aged 20–49 in low-and middle-income countries. Int J Obes 44(3):609–616
44. Hák T, Janoušková S, Moldan B (2016) Sustainable development goals: a need for relevant indicators. Ecol Indic 60:565–573
45. Webb P, Stordalen GA, Singh S, Wijesinha-Bettoni R, Shetty P, Lartey A (2018) Hunger and malnutrition in the 21st century. BMJ 361
46. Integrated Food Security Phase Classification (2025) IPC acute food insecurity classification. Available at: https://www.ipcinfo.org/ipcinfo-website/ipc-overview-and-classification-system/ipc-acute-food-insecurity-classification/en/. Accessed 2 March 2025
47. Myers SS, Smith MR, Guth S, Golden CD, Vaitla B, Mueller ND et al (2017) Climate change and global food systems: potential impacts on food security and undernutrition. Annu Rev Public Health 38(1):259–277
48. Moradi S, Mirzababaei A, Dadfarma A, Rezaei S, Mohammadi H, Jannat B, Mirzaei K (2019) Food insecurity and adult weight abnormality risk: a systematic review and meta-analysis. Eur J Nutr 58:45–61

49. Nettle D, Andrews C, Bateson M (2017) Food insecurity as a driver of obesity in humans: the insurance hypothesis. Behav Brain Sci 40:e105
50. Fonseca-Pérez D, Arteaga-Pazmiño C, Maza-Moscoso CP, Flores-Madrid S, Álvarez-Córdova L (2022) Food insecurity as a risk factor for sarcopenic obesity in older adults. Front Nutr 9:1040089
51. Pereira MH, Pereira ML, Campos GC, Molina MC (2022) Food insecurity and nutritional status among older adults: a systematic review. Nutr Rev 80(4):631–644
52. Abene JA, Tong J, Minuk J, Lindenfeldar G, Chen Y, Chao AM (2023) Food insecurity and binge eating: a systematic review and meta-analysis. Int J Eat Disord 56(7):1301–1322
53. Pourmotabbed A, Moradi S, Babaei A, Ghavami A, Mohammadi H, Jalili C et al (2020) Food insecurity and mental health: a systematic review and meta-analysis. Public Health Nutr 23(10):1778–1790
54. Meerman J, Aphane J (2012) Impact of high food prices on nutrition. Food and Agriculture Organization of the United Nations
55. Penne T, Goedemé T (2021) Can low-income households afford a healthy diet? Insufficient income as a driver of food insecurity in Europe. Food Policy 99:101978
56. Istituto Nazionale di Statistica (ISTAT) (2023) La povertà in Italia. Available at: https://www.istat.it/comunicato-stampa/la-poverta-in-italia-anno-2023/. Accessed 3 Mar 2025
57. Istituto Nazionale di Statistica (ISTAT) (2024) I numeri dell'insicurezza alimentare in Italia. Available at: https://www.cursa.it/wp-content/uploads/2024/10/ISTAT-Slide-2024.pdf. Accessed 3 Mar 2025
58. Ministero del Lavoro e delle Politiche Sociali (2023) Rapporto sui risultati aggregati—Indagine strutturata sui destinatari finali del Programma Operativo I FEAD anno 2022. Available at: https://www.lavoro.gov.it/temi-e-priorita/europa-e-fondi-europei/focus-on/fondo-di-aiuti-europei-agli-indigenti-Fead/Documents/FEAD-Report-Indagine-strutturata-2022.pdf. Accessed 3 Mar 2025
59. Okkio alla Salute (2014). https://www.epicentro.iss.it/okkioallasalute/pdf2015/sintesi_16gen.pdf
60. Zaçe D, Di Pietro ML, Reali L, De Waure C, Ricciardi W (2021) Prevalence, socioeconomic predictors and health correlates of food insecurity among Italian children-findings from a cross-sectional study. Food Secur 13:13–24
61. UNICEF (2023) The future of childhood in a changing world. Available at: https://www.unicef.org/reports/future-of-childhood-2023. Accessed 05 Mar 2025
62. Kerr JA, Patton GC, Cini KI, Abate YH, Abbas N, Abd Al Magied AH et al (2025) Global, regional, and national prevalence of child and adolescent overweight and obesity, 1990–2021, with forecasts to 2050: a forecasting study for the Global Burden of Disease Study 2021. Lancet 405(10481):785–812
63. World Health Organization (2006) Length/height-for-age child growth standards. Available at: https://www.who.int/tools/child-growth-standards/standards/length-height-for-age. Accessed 1 Feb 2025
64. World Health Organization (2006) Weight-for-age child growth standards. Available at: https://www.who.int/tools/child-growth-standards/standards/weight-for-age. Accessed 1 Feb 2025
65. World Health Organization (2006) Height-for-age reference data for 5 to 19 years. Available at: https://www.who.int/tools/growth-reference-data-for-5to19-years/indicators/height-for-age. Accessed 1 Feb 2025
66. Centers for Disease Control and Prevention (CDC, 2000) Clinical growth charts. Available at: https://www.cdc.gov/growthcharts/cdc-charts.htm. Accessed 4 Mar Apr 2025
67. Cole TJ, Lobstein T (2012) Extended international (IOTF) body mass index cut-offs for thinness, overweight and obesity. Pediatr Obes 7(4):284–294
68. Hampl SE, Hassink SG, Skinner AC, Armstrong SC, Barlow SE, Bolling CF et al (2023) Clinical practice guideline for the evaluation and treatment of children and adolescents with obesity. Pediatrics 151(2)

69. Jebeile H, Kelly AS, O'Malley G, Baur LA (2022) Obesity in children and adolescents: epidemiology, causes, assessment, and management. Lancet Diabetes Endocrinol 10(5):351–365
70. Wang Y, Lim H (2012) The global childhood obesity epidemic and the association between socio-economic status and childhood obesity. Int Rev Psychiatry 24(3):176–188
71. Brooks SJ, Feldman I, Schiöth HB, Titova OE (2021) Important gender differences in psychosomatic and school-related complaints in relation to adolescent weight status. Sci Rep 11(1):14147
72. Fan Y, Wang D, Wang W, Chen H, Lin J, Lin W et al (2024) Depression, anxiety, and development of obesity from childhood to young adulthood in China. J Psychosom Res 185:111867
73. Gallagher C, Waidyatillake N, Pirkis J, Lambert K, Cassim R, Dharmage S, Erbas B (2023) The long-term effects of childhood adiposity on depression and anxiety in adulthood: a systematic review. Obesity 31(9):2218–2228
74. Lindberg L, Hagman E, Danielsson P, Marcus C, Persson M (2020) Anxiety and depression in children and adolescents with obesity: a nationwide study in Sweden. BMC Med 18:1–9
75. Russell-Mayhew S, McVey G, Bardick A, Ireland A (2012) Mental health, wellness, and childhood overweight/obesity. J Obes 2012(1):281801
76. Renman C, Engström I, Silfverdal SA, Åman J (1999) Mental health and psychosocial characteristics in adolescent obesity: a population-based case-control study. Acta Paediatr 88(9):998–1003
77. Kang NR, Kwack YS (2020) An update on mental health problems and cognitive behavioral therapy in pediatric obesity. Pediatr Gastroenterol Hepatol Nutr 23(1):15
78. Cortese S, Moreira-Maia CR, St. Fleur D, Morcillo-Peñalver C, Rohde LA, Faraone SV (2016) Association between ADHD and obesity: a systematic review and meta-analysis. Am J Psychiatry 173(1):34–43
79. American Psychiatric Association (2022) Diagnostic and statistical manual of mental disorders (5th ed., text rev.). https://doi.org/10.1176/appi.books.9780890425787
80. Hirota T, King BH (2023) Autism spectrum disorder: a review. JAMA 329(2):157–168
81. Zablotsky B, Black LI, Blumberg SJ (2017) Estimated prevalence of children with diagnosed developmental disabilities in the United States, 2014–2016. NCHS Data Brief 291:1–8
82. Soltmann B, Neumann A, March S, Weinhold I, Häckl D, Kliemt R, Baum F, Romanos M, Schwarz J, von Peter S, Ignatyev Y, Arnold K, Swart E, Heinze M, Schmitt J, Pfennig A (2021) Multiperspective and multimethod evaluation of flexible and integrative psychiatric care models in Germany: study protocol of a prospective, controlled multicenter observational study (PsychCare). Front Psych 12
83. Autism Europe (2017). Il trattamento dei disturbi dello spettro autistico nei bambini e negli adolescenti. Available at: https://www.autismeurope.org/wp-content/uploads/2017/08/Il-trattamento-dei-disturbi-dello-spettro-autistico-nei-bambini-e-negli-adolescenti.pdf Accessed 29 Apr 2025
84. Sharma SR, Gonda X, Tarazi FI (2018) Autism Spectrum disorder: classification, diagnosis and therapy. Pharmacol Ther 190(1):91–104. [online]
85. Nimbley E, Golds L, Sharpe H, Gillespie-Smith K, Duffy F (2022) Sensory processing and eating behaviours in autism: a systematic review. Eur Eat Disord Rev. 30(5):538–559. https://doi.org/10.1002/erv.2920
86. Mørdre M, Ørbeck B, Hoel RE, Øvergaard KR (2024) Food selectivity in children and adolescents with autism spectrum disorders—a systematic literature review. Tidsskrift for Den norske legeforening. [online]
87. Page SD, Souders MC, Kral TVE, Chao AM, Pinto-Martin J (2021) Correlates of feeding difficulties among children with autism Spectrum disorder: a systematic review. J Autism Dev Disord 52
88. Matson JL, Sipes M, Fodstad JC, Fitzgerald ME (2011) Issues in the management of challenging behaviours of adults with autism spectrum disorder. CNS Drugs 25(7):597–606

89. Valenzuela-Zamora AF, Ramírez-Valenzuela DG, Ramos-Jiménez A (2022) Food selectivity and its implications associated with gastrointestinal disorders in children with autism spectrum disorders. Nutrients 14(13):2660
90. Rodrigues JVS, Poli MCF, Petrilli PH, Dornelles RCM, Turcio KH, Theodoro LH (2023) Food selectivity and neophobia in children with autism spectrum disorder and neurotypical development: a systematic review. Nutr Rev
91. Kuschner ES, Eisenberg IW, Orionzi B, Simmons WK, Kenworthy L, Martin A, Wallace GL (2015) A preliminary study of self-reported food selectivity in adolescents and young adults with autism spectrum disorder. Res Autism Spectr Disord 15-16:53–59
92. Maitin-Shepard M, O'Tierney-Ginn P, Kraneveld AD, Lyall K, Fallin D, Arora M, Fasano A, Mueller NT, Wang X, Caulfield LE, Dickerson AS, Diaz Heijtz R, Tarui T, Blumberg JB, Holingue C, Schmidt RJ, Garssen J, Almendinger K, Lin P-ID, Mozaffarian D (2024) Food, nutrition, and autism: from soil to fork. Am J Clin Nutr. [online] pp.S0002-9165(24)00443X
93. Conti MV, Breda C, Basilico S, Luzzi A, Voto L, Santero S, De Filippo G, Cena H (2023) Dietary recommendations to customize canteen menus according to the nutritional and sensory needs of individuals with autism spectrum disorder. Eat Weight Disord. Studies on Anorexia, Bulimia and Obesity 28:1
94. Marí-Bauset S, Zazpe I, Mari-Sanchis A, Llopis-González A, Morales-Suárez-Varela M (2013) Food selectivity in autism spectrum disorders. J Child Neurol 29(11):1554–1561
95. Bandini LG, Curtin C, Phillips S, Anderson SE, Maslin M, Must A (2016) Changes in food selectivity in children with autism Spectrum disorder. J Autism Dev Disord 47(2):439–446. [online]
96. Bayoumi SC, Halkett A, Miller M, Hinshaw SP (2025) Food selectivity and eating difficulties in adults with autism and/or ADHD. Autism
97. Li Y, Xie X, Lei X, Li Y, Lei X (2020) Global prevalence of obesity, overweight and underweight in children, adolescents and adults with autism spectrum disorder, attention-deficit hyperactivity disorder: a systematic review and meta-analysis. Obes Rev 21:12
98. Micai M, Fatta LM, Gila L, Caruso A, Salvitti T, Fulceri F, Ciaramella A, D'Amico R, Del Giovane C, Bertelli MO, Romano G, Schünemann HJ, Scattoni ML (2023) Prevalence of co-occurring conditions in children and adults with autism spectrum disorder: a systematic review and meta-analysis. Neurosci Biobehav Rev 155:105436
99. Sammels O, Karjalainen L, Dahlgren J, Wentz E (2022) Autism spectrum disorder and obesity in children: a systematic review and meta-analysis. Obes Facts 15:3
100. Esteban-Figuerola P, Canals J, Fernández-Cao JC, Arija Val V (2018) Differences in food consumption and nutritional intake between children with autism spectrum disorders and typically developing children: a meta-analysis. Autism 23(5):1079–1095
101. Coppola S, Nocerino R, Oglio F, Golia P, Falco MC, Riccio MP, Carucci L, Rea T, Simeone S, Garotti R, Marani N, Bravaccio C, Canani RB (2024) Adverse food reactions and alterations in nutritional status in children with autism spectrum disorders: results of the NAFRA project. Ital J Pediatr 50(1):228. [online]
102. Boltri M, Sapuppo W (2021) Anorexia nervosa and autism spectrum disorder: a systematic review. Psychiatry Res 306:114271
103. Nickel K, Maier S, Endres D, Joos A, Maier V, Tebartz van Elst L, Zeeck A (2019) Systematic review: overlap between eating, autism spectrum, and attention-deficit/hyperactivity disorder. Front Psych 10
104. Dreyer Gillette ML, Killian HJ, Fernandez C, Sweeney BR (2022) Treating obesity in children and adolescents with special healthcare needs. Current Obesity Reports
105. Dhaliwal KK, Orsso CE, Richard C, Haqq AM, Zwaigenbaum L (2019) Risk factors for unhealthy weight gain and obesity among children with autism spectrum disorder. Int J Mol Sci 20(13):3285
106. Sharp WG, Berry RC, Burrell L, Scahill L, McElhanon BO (2020) Scurvy as a sequela of avoidant-restrictive food intake disorder in autism: a systematic review. J Dev Behav Pediatr 41(5):397–405

107. Duignan E, Kenna PF, Watson R, Fitzsimon S, Brosnahan D (2015) Ophthalmic manifestations of vitamin a and D deficiency in two autistic teenagers: case reports and a review of the literature. Case Rep Ophthalmol 6(1):24–29
108. Pineles SL, Avery RA, Liu GT (2010) Vitamin B12 optic neuropathy in autism. Pediatrics 126(4):e967–e970
109. Zaenglein A, Martin A, Carlson L, Williams KE (2020) Pellagra secondary to selective eating in a child with autism. Pediatr Dermatol 37(4):698–700
110. García OP, Long KZ, Rosado JL (2009) Impact of micronutrient deficiencies on obesity. Nutr Rev 67(10):559–572. [online]
111. Nogueira-de-Almeida CA, de Araújo LA, da Ued VF, Contini AA, Nogueira-de-Almeida ME, Martinez EZ, Ferraz IS, Del Ciampo LA, Nogueira-de-Almeida CCJ, Fisberg M (2025) Nutritional factors and therapeutic interventions in autism Spectrum disorder: a narrative review. Children 12(2):202
112. Breda C, Santero S, Conti MV, Cena H (2024) Programmes to manage food selectivity in individuals with autism spectrum disorder. Nutr Res Rev, [online]:1–14
113. World Health Organization (2025a) Obesity. World Health Organization. Available at: https://www.who.int/health-topics/obesity#tab=tab_1. Accessed 29 Apr 2025 [online]
114. World Health Organization (2025b) Aging: global population. [online] www.who.int. Available at: https://www.who.int/news-room/questions-and-answers/item/population-ageing. Accessed 29 Apr 2025
115. United Nations (2024) Global issues: aging. [online] United Nations. Available at: https://www.un.org/en/global-issues/ageing. Accessed 29 Apr 2025
116. Beard JR, Bloom DE (2015) Toward a comprehensive public health response to population aging. Lancet 385(9968):658–661
117. Scott AJ (2021) The longevity society. Lancet Healthy Longev 2(12):e820–e827
118. Eurostat (2024) Aging Europe—introduction—statistics explained. [online] Available at: https://ec.europa.eu/eurostat/statistics-explained/index.php?title=Ageing_Europe_-_introduction. Accessed 29 Apr 2025
119. Calder PC, Bosco N, Bourdet-Sicard R, Capuron L, Delzenne N, Doré J, Franceschi C, Lehtinen MJ, Recker T, Salvioli S, Visioli F (2017) Health relevance of the modification of low grade inflammation in aging (inflammageing) and the role of nutrition. Aging Res Rev 40:95–119. [online]
120. Raymond JL, Morrow K (2022) Krause and Mahan's food and the nutrition care process, 16e, E-book, 16th edn. Elsevier Health Sciences
121. López-Otín C, Blasco MA, Partridge L, Serrano M, Kroemer G (2023) Hallmarks of aging: an expanding universe. Cell 186(2):243–278. [online]
122. Cruz-Jentoft AJ, Sayer AA (2019) Sarcopenia. Lancet 393(10191):2636–2646. Corrigendum to 'Sarcopenia' [Lancet (2019) 393:2590]
123. Neeland IJ, Ross R, Després J-P, Matsuzawa Y, Yamashita S, Shai I, Seidell J, Magni P, Santos RD, Arsenault B, Cuevas A, Hu FB, Griffin B, Zambon A, Barter P, Fruchart J-C, Eckel RH (2019) Visceral and ectopic fat, atherosclerosis, and cardiometabolic disease: a position statement. Lancet Diabetes Endocrinol 7(9):715–725. [online]
124. Dent E, Wright ORL, Woo J, Hoogendijk EO (2023) Malnutrition in older adults. Lancet 401:10380
125. World Health Organization (2017) Integrated care for older people Guidelines on community-level interventions to manage declines in intrinsic capacity. [online] Available at: https://iris.who.int/bitstream/handle/10665/258981/9789241550109-eng.pdf?sequence=1. Accessed 29 Apr 2025
126. Franceschi C, Bonafè M, Valensin S, Olivieri F, De Luca M, In Ottaviani E, addition, De Benedictis, G. (2006) Inflamm-aging: an evolutionary perspective on immunosenescence. Ann N Y Acad Sci 908(1):244–254

127. Franceschi C, Garagnani P, Parini P, Giuliani C, Santoro A (2018) Inflammaging: a new immune–metabolic viewpoint for age-related diseases. Nat Rev Endocrinol 14(10):576–590. [online]
128. AJMC (2006) Vulnerable populations: who are they? [online] AJMC. Available at: https://www.ajmc.com/view/nov06-2390ps348-s352. Accessed 29 Apr 2025
129. Volkert D, Kiesswetter E, Cederholm T, Donini LM, Eglseer D, Norman K, Schneider SM, Ströbele-Benschop N, Torbahn G, Wirth R, Visser M (2019a) Development of a model on determinants of malnutrition in aged persons: a MaNuEL project. Gerontol Geriatr Med., [online] 5,:233372141985843
130. Volkert D, Visser M, Corish CA, Geisler C, de Groot L, Cruz-Jentoft AJ, Lohrmann C, O'Connor EM, Schindler K, de van der Schueren MAE (2019b) Joint action malnutrition in the elderly (MaNuEL) knowledge hub: summary of project findings. Eur Geriatr Med 11(1):169–177
131. Nature Food (2023) The triple burden of malnutrition. Nat Food 4(11):925–925. [online]
132. Cederholm T, Jensen GL, Correia MITD, Gonzalez MC, Fukushima R, Higashiguchi T, Baptista G, Barazzoni R, Blaauw R, Coats A, Crivelli A, Evans DC, Gramlich L, Fuchs-Tarlovsky V, Keller H, Llido L, Malone A, Mogensen KM, Morley JE, Muscaritoli M (2019) GLIM criteria for the diagnosis of malnutrition—a consensus report from the global clinical nutrition community. Clin Nutr (Edinburgh, Scotland)., [online] 38(1):1–9
133. Kyle U (2004) Bioelectrical impedance analysis? Part I: review of principles and methods. Clin Nutr 23(5):1226–1243. [online]
134. Kyle UG, Bosaeus I, De Lorenzo AD, Deurenberg P, Elia M, Manuel Gómez J, Lilienthal Heitmann B, Kent-Smith L, Melchior J-C, Pirlich M, Scharfetter H, Schols MWJA, Pichard C (2004) Bioelectrical impedance analysis—part II: utilization in clinical practice. Clin Nutr 23(6):1430–1453. [online]
135. Sergi G, De Rui M, Veronese N, Bolzetta F, Berton L, Carraro S, Bano G, Coin A, Manzato E, Perissinotto E (2015) Assessing appendicular skeletal muscle mass with bioelectrical impedance analysis in free-living Caucasian older adults. Clin Nutr 34(4):667–673
136. Barazzoni R, Jensen GL, Correia MITD, Gonzalez MC, Higashiguchi T, Shi HP, Bischoff SC, Boirie Y, Carrasco F, Cruz-Jentoft A, Fuchs-Tarlovsky V, Fukushima R, Heymsfield S, Mourtzakis M, Muscaritoli M, Norman K, Nyulasi I, Pisprasert V, Prado C, de van der Schuren M (2022) Guidance for assessment of the muscle mass phenotypic criterion for the Global Leadership Initiative on Malnutrition (GLIM) diagnosis of malnutrition. Clin Nutr., [online] 41(6):1425–1433
137. Cederholm T, Jensen GL, Ballesteros-Pomar MD et al (2024) Guidance for assessment of the inflammation etiologic criterion for the GLIM diagnosis of malnutrition: a modified Delphi approach. Clin Nutr 43(5):1025–1032. https://doi.org/10.1016/j.clnu.2023.11.026
138. Crichton M, Craven D, Mackay H, Marx W, de van der Schueren M, Marshall S (2019) A systematic review, meta-analysis and meta-regression of the prevalence of protein-energy malnutrition: associations with geographical region and sex. Age and Aging 48(1):38–48. [online]
139. Muscaritoli M, Amabile MI, Gallicchio C et al (2025) Prevalence of malnutrition and its impact on clinical outcomes in internal medicine wards: the SIMI-NUTRO study. Eur J Intern Med 138:83–90
140. Alves LF, de Jesus JDS, Britto VNM, de Jesus SA, Santos GS, de Oliveira CC (2023) GLIM criteria to identify malnutrition in patients in hospital settings: a systematic review. J Parenter Enter Nutr 47(6):702–709
141. Bian W, Li Y, Wang Y, Chang L, Deng L, Li Y, Jiang H, Zhou P (2023) Prevalence of malnutrition based on global leadership initiative in malnutrition criteria for completeness of diagnosis and future risk of malnutrition based on current malnutrition diagnosis: systematic review and meta-analysis. Front Nutr 10
142. Lingvay I, Cohen RV, Roux CWL, Sumithran P (2024) Obesity in adults. Lancet 404(10456):972–987

143. World Health Organization (2010a) A healthy lifestyle—WHO Recommendations. [online] World Health Organization. Available at: https://www.who.int/europe/news-room/fact-sheets/item/a-healthy-lifestyle%2D%2D-who-recommendations. Accessed 29 Apr 2025
144. World Health Organization (2010b) Healthy workplaces: a model for action: for employers, workers, policy-makers and practitioners. In Healthy workplaces: a model for action: for employers, workers, policy-makers and practitioners
145. World Health Organization (2008) Waist circumference and waist-hip ratio. [online] Available at: https://iris.who.int/bitstream/handle/10665/44583/9789241501491_eng.pdf?sequence=1. Accessed 29 Apr 2025
146. International Diabetes Federation (2006) Metabolic syndrome. [online] Available at: https://idf.org/media/uploads/2023/05/attachments-30.pdf. Accessed 29 Apr 2025
147. National Institute for Health and Care Excellence (2022) Keep the size of your waist to less than half of your height, NICE recommends. [online] NICE. Available at: https://www.nice.org.uk/news/articles/keep-the-size-of-your-waist-to-less-than-half-of-your-height-nice%2D%2Drecommends. Accessed 29 Apr 2025
148. Government Office for Science (2007) Foresight obesity system map. UK Government. https://assets.publishing.service.gov.uk/media/5a7c398f40f0b674ed20faac/obesity-map-full-hi-res.pdf
149. Khaleghi AA, Salari N, Darvishi N, Bokaee S, Jafari S, Hemmati M, Mohammadi M (2025) Global prevalence of obesity in the older adults: a meta-analysis. Public Health Pract., [online] 9:100585
150. EpiCentro (2023) EpiCentro. [online] Epicentro.iss.it. Available at: https://www.epicentro.iss.it/passi-argento/dati/obesita. Accessed 29 Apr 2025
151. Berger MM, Shenkin A, Schweinlin A, Amrein K, Augsburger M, Biesalski H-K, Bischoff SC, Casaer MP, Gundogan K, Lepp H-L, Man M, Pietka GM, Pironi L, Rezzi S, Cuerda C (2024) Corrigendum to 'ESPEN micronutrient guideline' [Clin Nutr 41 (2022) 1357–1424/YCLNU5151]. Clin Nutr 43(4):1024–1024
152. Ter Borg S, Verlaan S, Hemsworth J, Mijnarends DM, Schols JMGA, Luiking YC, de Groot LCPGM (2015) Micronutrient intakes and potential inadequacies of community-dwelling older adults: a systematic review. Br J Nutr 113(8):1195–1206
153. Passarelli S, Free CM, Shepon A, Beal T, Batis C, Golden CD (2024) Global estimation of dietary micronutrient inadequacies: a modeling analysis. Lancet Glob Health 12:10
154. Castiglione D, Platania A, Conti A, Falla M, D'Urso M, Marranzano M (2018) Dietary micronutrient and mineral intake in the Mediterranean healthy eating, aging, and lifestyle (MEAL) study. Antioxidants 7(7):79. [online]
155. Pradelli L, Zaniolo O, Sanfilippo A, Lezo A, Riso S, Zanetti M (2023) Prevalence and economic cost of malnutrition in Italy: a systematic review and metanalysis from the Italian Society of Artificial Nutrition and Metabolism (SINPE). Nutrition:111943
156. OnFoods (2024) Sustainable eating patterns to limit malnutrition in older adults | OnFoods. [online] Available at: https://www.onfoods.it/research-projects/sustainable-eating-patterns-limit-malnutrition-older-adults. Accessed 29 Apr 2025
157. Schuetz P, Seres D, Lobo DN, Gomes F, Kaegi-Braun N, Stanga Z (2021) Management of disease-related malnutrition for patients being treated in hospital. Lancet 398(10314):1927–1938
158. Butterworth CE Jr (1994) The skeleton in the hospital closet. Nutrition 10(5):442
159. Marshall S (2018) Why is the skeleton still in the hospital closet? A look at the complex etiology of protein-energy malnutrition and its implications for the nutrition care team. J Nutr Health Aging 22(1):26–29
160. Sharma Y, Thompson C, Shahi R, Hakendorf P, Miller M (2017) Malnutrition in acutely unwell hospitalized elderly—'the skeletons are still rattling in the hospital closet'. J Nutr Health Aging 21(10):1210–1215
161. Dent E, Hoogendijk EO, Visvanathan R, Wright ORL (2019a) Malnutrition screening and assessment in hospitalized older people: a review. J Nutr Health Aging 23(5):431–441. [online]

162. Dent E, Morley JE, Cruz-Jentoft AJ, Woodhouse L, Rodríguez-Mañas L, Fried LP, Woo J, Aprahamian I, Sanford A, Lundy J, Landi F, Beilby J, Martin FC, Bauer JM, Ferrucci L, Merchant RA, Dong B, Arai H, Hoogendijk EO, Won CW (2019b) Physical frailty: ICFSR international clinical practice guidelines for identification and management. J Nutr Health Aging 23(9):771–787
163. Díaz G, Correia MITD, Gonzalez MC, Reyes M (2023) The global leadership initiative on malnutrition criteria for the diagnosis of malnutrition in patients admitted to the intensive care unit: a systematic review and meta-analysis. Clin Nutr 42(2):182–189. [online]
164. Muñoz Fernandez SS, Garcez FB, de Alencar JCG, Cederholm T, Aprahamian I, Morley JE, de Souza HP, Avelino da Silva TJ, Ribeiro SML (2021) Applicability of the GLIM criteria for the diagnosis of malnutrition in older adults in the emergency ward: a pilot validation study. Clin Nutr 40(11):5447–5456. [online]
165. Muscaritoli M, Arends J, Bachmann P, Baracos V, Barthelemy N, Bertz H, Bozzetti F, Hütterer E, Isenring E, Kaasa S, Krznaric Z, Laird B, Larsson M, Laviano A, Mühlebach S, Oldervoll L, Ravasco P, Solheim TS, Strasser F, de van der Schueren M (2021) ESPEN practical guideline: Clinical Nutrition in cancer. Clin Nutr, [online] 40:5
166. Bischoff SC, Bager P, Escher J, Forbes A, Hébuterne X, Hvas CL, Joly F, Klek S, Krznaric Z, Ockenga J, Schneider S, Shamir R, Stardelova K, Bender DV, Wierdsma N, Weimann A (2023) ESPEN guideline on clinical nutrition in inflammatory bowel disease. Clin Nutr 42:3
167. Deng M, Lu Y, Zhang Q, Bian Y, Zhou X, Hou G (2023) Global prevalence of malnutrition in patients with chronic obstructive pulmonary disease: systemic review and meta-analysis. Clin Nutr 42(6):848–858
168. Weimann A, Braga M, Carli F, Higashiguchi T, Hübner M, Klek S, Laviano A, Ljungqvist O, Lobo DN, Martindale RG, Waitzberg D, Bischoff SC, Singer P (2021) ESPEN practical guideline: clinical nutrition in surgery. Clin Nutr 40(7):4745–4761. [online]
169. Zhang T, Qin J, Guo J, Dong J, Chen J, Ma Y, Han L (2024) Prevalence and influencing factors of malnutrition in diabetic patients: a systematic review and meta-analysis. J Diabetes 16:10
170. Sabatino A, Fiaccadori E, Barazzoni R, Carrero JJ, Cupisti A, De Waele E, Jonckheer J, Cuerda C, Bischoff SC (2024) Corrigendum to 'ESPEN practical guideline on clinical nutrition in hospitalized patients with acute or chronic kidney disease' [Clin Nutr 43 (2024) 2238–2254]. Clin Nutr 43(10):2394–2398
171. Burgos R, Bretón I, Cereda E, Desport JC, Dziewas R, Genton L, Gomes F, Jésus P, Leischker A, Muscaritoli M, Poulia K-A, Preiser JC, Van der Marck M, Wirth R, Singer P, Bischoff SC (2018) ESPEN guideline clinical nutrition in neurology. Clin Nutr 37(1):354–396. [online]
172. Volkert D, Beck AM, Faxén-Irving G, Frühwald T, Hooper L, Keller H, Porter J, Rothenberg E, Suominen M, Wirth R, Chourdakis M (2024) ESPEN guideline on nutrition and hydration in dementia—update 2024. Clin Nutr 43:6
173. Fukushima R, Compher CW, Correia MITD, Gonzalez MC, McKeever L, Nakamura K, Lee ZY, Patel JJ, Singer P, Stoppe C, Ayala JC, Barazzoni R, Berger MM, Cederholm T, Chittawatanarat K, Cotoia A, Lopez-Delgado JC, Earthman CP, Elke G, Hartl W (2025) Recognizing malnutrition in adults with critical illness: guidance statements from the global leadership initiative on malnutrition. Clin Nutr
174. Milanez DSJ, Razzera EL, Lima J, Silva FM (2023) Feasibility and criterion validity of the GLIM criteria in the critically ill: a prospective cohort study. J Parenter Enter Nutr 47(6):754–765
175. Gersely GD, Camila R, Del G, Ferreira W, Mara L, Santos BC, Manuela M, Junior S, Isabel M, Waitzberg DL, Ozorio GA (2024) GLIM criteria validation and reliability in critically ill patients with cancer: a prospective study. J Parenter Enter Nutr 48(6):726–734
176. O'Keeffe M, Kelly M, O'Herlihy E, O'Toole PW, Kearney PM, Timmons S, O'Shea E, Stanton C, Hickson M, Rolland Y, Sulmont Rossé C, Issanchou S, Maitre I, Stelmach-Mardas M, Nagel G, Flechtner-Mors M, Wolters M, Hebestreit A, De Groot LCPGM, van de Rest O (2019) Potentially modifiable determinants of malnutrition in older adults: a systematic review. Clin Nutr 38(6):2477–2498. [online]

177. Cass AR, Charlton KE (2022) Prevalence of hospital-acquired malnutrition and modifiable determinants of nutritional deterioration during inpatient admissions: a systematic review of the evidence. J Hum Nutr Diet 35(6):1043–1058
178. Laur C, McCullough J, Davidson B, Keller H (2015) Becoming food aware in hospital: a narrative review to advance the culture of nutrition Care in Hospitals. Healthcare 3(2):393–407. [online]
179. Porter J, Haines TP, Truby H (2017) The efficacy of protected mealtimes in hospitalized patients: a stepped wedge cluster randomized controlled trial. BMC Med 15:1
180. Young AM, Mudge AM, Banks MD, Ross LJ, Daniels L (2013) Encouraging, assisting and time to EAT: improved nutritional intake for older medical patients receiving protected mealtimes and/or additional nursing feeding assistance. Clin Nutr 32(4):543–549
181. Young AM, Banks MD, Mudge AM (2018) Improving nutrition care and intake for older hospital patients through system-level dietary and mealtime interventions. Clin Nutr ESPEN 24(1):140–147
182. BAPEN (2023) Who is at risk of malnutrition? [online] BAPEN. Available at: https://www.bapen.org.uk/malnutrition/introduction-to-malnutrition/who-is-at-risk-of-malnutrition/. Accessed 29 Apr 2025
183. Thibault R, Abbasoglu O, Ioannou E, Meija L, Ottens-Oussoren K, Pichard C, Rothenberg E, Rubin D, Siljamäki-Ojansuu U, Vaillant M-F, Bischoff SC (2021) ESPEN guideline on hospital nutrition. Clin Nutr., [online] 40:12
184. Volkert D, Beck AM, Cederholm T, Cruz-Jentoft A, Hooper L, Kiesswetter E, Maggio M, Raynaud-Simon A, Sieber C, Sobotka L, van Asselt D, Wirth R, Bischoff SC (2022) ESPEN practical guideline: clinical nutrition and hydration in geriatrics. Clin Nutr 41(4):958–989. [online]
185. SINPE, Società Italiana di Nutrizione Parenterale ed Enterale (2024) Screening nutrizionali. [online] Available at: https://www.sinpe.org/area-servizi/screening-nutrizionali.html. Accessed 29 Apr 2025
186. Schuetz P, Fehr R, Baechli V, Geiser M, Deiss M, Gomes F, Kutz A, Tribolet P, Bregenzer T, Braun N, Hoess C, Pavlicek V, Schmid S, Bilz S, Sigrist S, Brändle M, Benz C, Henzen C, Mattmann S, Thomann R (2019) Individualized nutritional support in medical inpatients at nutritional risk: a randomized clinical trial. Lancet 393(10188):2312–2321
187. European Society for Clinical Nutrition and Metabolism ESPEN (2019) ESPEN Guidelines. [online] ESPEN. Available at: https://www.espen.org/guidelines-home/espen-guidelines. Accessed 29 Apr 2025
188. Wong A, Huang Y, Sowa PM, Banks MD, Bauer JD (2023) An umbrella review and meta-analysis of interventions, excluding enteral and parenteral nutrition, initiated in the hospital for adults with or at risk of malnutrition. Am J Clin Nutr., [online] 118:3
189. Brunner S, Mayer H, Qin H, Breidert M, Dietrich M, Müller Staub M (2021) Interventions to optimize nutrition in older people in hospitals and long-term care: umbrella review. Scand J Caring Sci 36:3
190. Correa-Pérez A, Abraha I, Cherubini A, Collinson A, Dardevet D, de Groot LCPGM, de van der Schueren MAE, Hebestreit A, Hickson M, Jaramillo-Hidalgo J, Lozano-Montoya I, O'Mahony D, Soiza RL, Visser M, Volkert D, Wolters M, Jentoft AJC (2019) Efficacy of non-pharmacological interventions to treat malnutrition in older persons: a systematic review and meta-analysis. The SENATOR project ONTOP series and MaNuEL knowledge hub project. Aging Res Rev 49:27–48
191. Poscia A, Milovanovic S, La Milia DI, Duplaga M, Grysztar M, Landi F, Moscato U, Magnavita N, Collamati A, Ricciardi W (2017) Effectiveness of nutritional interventions addressed to elderly persons: umbrella systematic review with meta-analysis. Eur J Pub Health 28(2):275–283
192. Rasmussen NML, Belqaid K, Lugnet K, Nielsen AL, Rasmussen HH, Beck AM (2018) Effectiveness of multidisciplinary nutritional support in older hospitalized patients: a systematic review and meta-analyses. Clin Nutr ESPEN 27:44–52

193. Hersberger L, Dietz A, Bürgler H, Bargetzi A, Bargetzi L, Kägi-Braun N, Tribolet P, Gomes F, Hoess C, Pavlicek V, Bilz S, Sigrist S, Brändle M, Henzen C, Thomann R, Rutishauser J, Aujesky D, Rodondi N, Donzé J, Stanga Z (2021) Individualized nutritional support for hospitalized patients with chronic heart failure. J Am Coll Cardiol 77(18):2307–2319
194. Bargetzi L, Brack C, Herrmann J, Bargetzi A, Hersberger L, Bargetzi M, Kaegi-Braun N, Tribolet P, Gomes F, Hoess C, Pavlicek V, Bilz S, Sigrist S, Brändle M, Henzen C, Thomann R, Rutishauser J, Aujesky D, Rodondi N, Donzé J (2021a) Nutritional support during the hospital stay reduces mortality in patients with different types of cancers: secondary analysis of a prospective randomized trial. Ann Oncol 32(8):1025–1033
195. Bargetzi A, Emmenegger N, Wildisen S, Nickler M, Bargetzi L, Hersberger L, Segerer S, Kaegi-Braun N, Tribolet P, Gomes F, Hoess C, Pavlicek V, Bilz S, Sigrist S, Brändle M, Henzen C, Thomann R, Rutishauser J, Aujesky D, Rodondi N (2021b) Admission kidney function is a strong predictor for the response to nutritional support in patients at nutritional risk. Clin Nutr 40(5):2762–2771
196. Keller B, Wunderle C, Tribolet P, Stanga Z, Kaegi-Braun N, Mueller B, Schuetz P (2024) Nutritional support in hospitalized patients with diabetes and risk for malnutrition: a secondary analysis of an investigator-initiated, Swiss, randomized controlled multicenter trial. BMJ Open 14(8):e084754–e084754
197. MNA elderly (2021). Available at: https://www.mna-elderly.com/sites/default/files/2021-10/mna-mini-italian.pdf Accessed 29 Apr 2025
198. Kaiser MJ, Bauer JM, Ramsch C, Uter W, Guigoz Y, Cederholm T, Thomas DR, Anthony P, Charlton KE, Maggio M, Tsai AC, Grathwohl D, Vellas B, Sieber CC (2009) Validation of the mini nutritional assessment short-form (MNA®-SF): a practical tool for identification of nutritional status. J Nutr Health Aging 13(9):782–788. [online]
199. Kaiser MJ, Bauer JM, Uter W, Donini LM, Stange I, Volkert D, Diekmann R, Drey M, Bollwein J, Tempera S, Guerra A, Ricciardi LM, Sieber CC (2011) Prospective validation of the modified mini nutritional assessment short-forms in the community, nursing home, and rehabilitation setting. J Am Geriatr Soc 59(11):2124–2128
200. Stratton RJ, Hackston A, Longmore D, Dixon R, Price S, Stroud M, King C, Elia M (2004) Malnutrition in hospital outpatients and inpatients: prevalence, concurrent validity and ease of use of the 'malnutrition universal screening tool' ('MUST') for adults. Br J Nutr 92(5):799–808. [online]
201. MNA-SF (MNA) (2021) https://www.mna-elderly.com/sites/default/files/2021-10/mna-mini-english.pdf
202. MUST (BAPEN). https://www.bapen.org.uk/pdfs/must/must_full.pdf
203. Ministero della Salute (2020) Direzione Generale per l'igiene, la Sicurezza degli Alimenti e la Nutrizione. Miglioramento della salute dell'anziano per gli aspetti nutrizionali (con particolare riguardo alla malnutrizione per difetto). Available at: https://www.salute.gov.it/imgs/C_17_pubblicazioni_3074_allegato.pdf Accessed 29 Apr 2025
204. Regione Lombardia (2024) DELIBERA N.1812- (DL) Attivazione Del Percorso Di Screening Nutrizionale Nelle Strutture Del Ssr E Domiciliare. Available online at: https://www.regione.lombardia.it/wps/portal/istituzionale/HP/istituzione/Giunta/sedute-delibere-giunta-regionale/DettaglioDelibere/delibera-1812-legislatura-12 Accessed 29 Apr 2025
205. D.D.G. n. 120293, settembre 2025, Approvazione degli «Indirizzi regionali per l'effettuazione dello screening nutrizionale e la gestione del rischio di malnutrizione». Regione Lombardia, Available at: https://omceomi.it/wp-content/uploads/2025/09/2-Approvazione-indirizzi-regionali-per-effettuazione-screening-nutrizionale-e-gestione-rischio-di-malnutrizione-D.G.-Welfare-D.d.g.-3-settembre-2025-n.-12029-BURL-9.9.2025-1.pdf. Accessed 29 Apr 2025
206. Mattavelli E, Verduci E, Mascheroni A et al (2025) Toward a pragmatic multidisciplinary management of nutritional risk in hospitalized patients: initiatives and proposals of the clinical nutrition network of lombardy region. Nutrients 17(9):1472. Published 2025 Apr 27. https://doi.org/10.3390/nu17091472
207. European Network for Workplace Health Promotion (n.d.) Luxembourg declaration on workplace health promotion. Available at: http://www.enwhp.org/

208. Beresford SA, Thompson B, Feng Z, Christianson A, McLerran D, Patrick DL (2001) Seattle 5 a day worksite program to increase fruit and vegetable consumption. Prev Med 32(3):230–238
209. Sorensen G, Linnan L, Hunt MK (2004) Worksite-based research and initiatives to increase fruit and vegetable consumption. Prev Med 39:94–100
210. World Health Organization (2016) Action plan for the prevention and control of noncommunicable diseases in the WHO European Region. In Action plan for the prevention and control of noncommunicable diseases in the WHO European Region. Available online at: https://iris.who.int/bitstream/handle/10665/341522/WHO-EURO-2016-2582-42338-58618-eng.pdf?sequence=1. Accessed 5 May 2025
211. Cahalin LP, Kaminsky L, Lavie CJ, Briggs P, Cahalin BL, Myers J et al (2015a) Development and implementation of worksite health and wellness programs: a focus on non-communicable disease. Prog Cardiovasc Dis 58(1):94–101
212. Cahalin LP, Kaminsky L, Lavie CJ, Briggs P, Cahalin BL, Myers J, Forman DE, Patel MJ, Pinkstaff SO, Arena R (2015b) Development and implementation of worksite health and wellness programs: a focus on non-communicable disease. Prog Cardiovasc Dis 58(1):94–101
213. Amani R, Gill T (2011) Shiftworking, nutrition and obesity: implications for workforce health—a systematic review. Obes Res Clin Pract 5:61
214. Virgara R, Singh B, O'Connor E, Szeto K, Merkx Z, Rees C, Gilson N, Maher C (2024) Keep on truckin': how effective are health behaviour interventions on truck drivers' health? A systematic review and meta-analysis. BMC Public Health 24:1
215. Safi F, Aniserowicz AM, Colquhoun H, Stier J, Nowrouzi-Kia B (2022) Impact of eating disorders on paid or unpaid work participation and performance: a systematic review and meta-analysis protocol. J Eat Disord 10(1):7
216. Sowah D, Fan X, Dennett L, Hagtvedt R, Straube S (2017) Vitamin D levels and deficiency with different occupations: a systematic review. BMC Public Health 17(1):519. Published 2017 Jun 22. https://doi.org/10.1186/s12889-017-4436-z
217. Shahudin NN, Sameeha MJ, Mat Ludin AF, Manaf ZA, Chin KY, Jamil NA (2020) Barriers towards sun exposure and strategies to overcome these barriers in female indoor workers with insufficient vitamin D: a qualitative approach. Nutrients 12(10):2994

Open Access This chapter is licensed under the terms of the Creative Commons Attribution-NonCommercial-NoDerivatives 4.0 International License (http://creativecommons.org/licenses/by-nc-nd/4.0/), which permits any noncommercial use, sharing, distribution and reproduction in any medium or format, as long as you give appropriate credit to the original author(s) and the source, provide a link to the Creative Commons license and indicate if you modified the licensed material. You do not have permission under this license to share adapted material derived from this chapter or parts of it.

The images or other third party material in this chapter are included in the chapter's Creative Commons license, unless indicated otherwise in a credit line to the material. If material is not included in the chapter's Creative Commons license and your intended use is not permitted by statutory regulation or exceeds the permitted use, you will need to obtain permission directly from the copyright holder.

Chapter 3
Economic Impacts of Malnutrition

Beatrice Re, G. Viroli, S. Santero, and Pietro Previtali

Economic Effects of the Triple Burden of Malnutrition

Nutrition economics has emerged as a distinct subdiscipline within health economics, formally introduced in the early 2010s by a multidisciplinary group of experts. They defined it as the study of health and economic outcomes related to nutrition for the benefit of society [1]. Nutrition is widely recognized as a key modifiable determinant of health and disease. This was underscored during the 2011 United Nations General Assembly on the prevention and control of NCDs, where the critical importance of implementing preventative health strategies was affirmed.

Between 1990 and 2017, premature average mortality from NCDs declined in all WHO regions and most countries. However, only the Western Pacific, European regions, and 25 mostly high-income countries are on track to meet SDG target 3.4. Since 2017, progress has slowed globally. High mortality rates remain concentrated in LMICs, particularly in Southeast Asia, the Eastern Mediterranean, and African countries, with the greatest reductions by 2017, which had already reached low mortality levels [2].

Concurrently, healthcare spending continues to rise due to factors such as birth rates, income, inflation, and 'excess growth'—mainly driven by medical technology and increased service demand. This growth increases healthcare's share of GDP, challenging fiscal sustainability [3].

B. Re (✉) · P. Previtali
Department of Economics and Management, University of Pavia, Pavia, Italy
e-mail: beatrice.re@unipv.it; pietro.previtali@unipv.it

G. Viroli · S. Santero
Department of Public Health, Experimental and Forensic Medicine, University of Pavia, Pavia, Italy
e-mail: giulia.viroli@unipv.it; sara.santero@unipv.it

© The Author(s) 2026

B. Re, P. Previtali (eds.), *Tackling Malnutrition in a Complex World*,
SpringerBriefs in Business, https://doi.org/10.1007/978-3-032-26438-1_3

In this context, nutrition economics plays a critical role in informing and prioritizing cost-effective, nutrition-based preventative interventions aimed at improving population health and ensuring the sustainability of healthcare systems.

Economic Threshold for Affording a Healthy Diet

Healthy diets are fundamental to achieving food security and enhancing nutritional outcomes. They consist of diverse, safe, and nutrient-rich foods that meet the body's energy and nutritional needs to maintain a healthy and active lifestyle. These diets prioritize unprocessed or minimally processed foods, which are balanced across food groups, while limiting the intake of highly processed foods and beverages. The key components include whole grains; legumes; nuts; a wide variety of fruits and vegetables; optional moderate consumption of eggs; dairy, poultry, and fish; and limited intake of red meat. Maintaining a healthy diet across the life course is vital for preventing all forms of malnutrition—including stunting, wasting, micronutrient deficiencies, and obesity—and for lowering the risk of NCDs such as cardiovascular disease, diabetes, and certain cancers [4].

Globally, the cost of a healthy diet increased by 4.3% in 2021 compared with that in 2020 and by 6.7% relative to prepandemic levels in 2019 [5]. This rise is attributed primarily to broad inflationary pressures during 2020 and 2021, partly stemming from the ongoing impacts of the COVID-19 pandemic. Data show that nearly 3.2 billion people were unable to afford a healthy diet in 2020 (ibid). This number slightly decreased in 2021, reaching over 3.1 billion people—equivalent to 42% of the global population. These data signal an increase of 134 million individuals compared with 2019, prior to the onset of the COVID-19 pandemic. This rise is attributed to the growing cost of healthy diets, which, in many countries, coincided with a reduction in disposable income. Indeed, in 2022, food prices continued to increase compared with those in the 2017–2021 period, leading to a rise in the average cost of a healthy diet both globally and across all regions [6].

Economic Impact of Childhood Malnutrition on "Adults of the Future"

Childhood obesity is one of the most pressing and concerning health issues worldwide. Over the past 40 years, global childhood obesity rates have increased more than eightfold. In 2020, approximately 39 million children under the age of 5 and 150 million children aged 5–19 years were overweight or with obesity. These figures are projected to rise to 40 million and 254 million, respectively, by 2030 [7]. The systematic review by Ling et al. [8] does not offer a reassuring scenario, as it shows that the annual total medical costs attributable to childhood overweight and obesity increased by $237.55 per capita. These costs were broken down as follows: $56.52 for nonhospital healthcare, $14.27 for outpatient visits, $46.38 for medication, and $1975.06 for hospitalization. Additionally, the length of hospital stays increased by an average of 0.28 days. By 2050, the annual direct and indirect costs are projected to reach $13.62 billion and $49.02 billion, respectively. During childhood, direct medical expenditures were higher for males than for females, but this trend reversed in adulthood, with females incurring higher expenditures. Childhood overweight and obesity pose health and economic burdens for both individuals and society. The **immediate economic impacts** may include healthcare expenses such as costs for prescription medications, emergency room visits, and both outpatient and inpatient care. Moreover, the **indirect economic effects** can extend to the labor market, including increased job absenteeism and reduced productivity among caregivers tending to ill children [9]. Additionally, excess weight in childhood often continues into adulthood, leading to elevated long-term costs due to obesity-related health conditions and their treatment [10]. Children with health issues may achieve lower levels of education, and education can influence the impact of other social determinants of health, such as income. These broader patterns linking education and health may also extend to obesity [11]. Recently, Segal et al. [12] suggested that obesity-related health conditions negatively impact educational outcomes, particularly when exposure is measured at age 12 or older. Interestingly, adverse effects are more pronounced and frequent among girls than boys.

Economic Impact of Older Adults' Malnutrition

Malnutrition is highly prevalent among older adults and imposes a significant burden on healthcare, social, and aged-care systems. Age-related physiological changes, limited access to nutritious food, and the presence of comorbidities contribute to this vulnerability. Clinical guidelines advocate for routine malnutrition screening in all older adults, followed by comprehensive nutritional assessment and personalized nutritional interventions for those who screen positive [13].

Specifically, the current European Society for Clinical Nutrition and Metabolism (ESPEN) guidelines [14] suggest that older persons with malnutrition or at risk of malnutrition with chronic conditions should be offered oral nutritional supplements (ONS) when dietary counseling and food fortification are not sufficient to increase dietary intake and reach nutritional goals. With reference to the hospital setting, a secondary analysis of the NOURISH study (multicenter, randomized controlled trial) analyzed the effectiveness and cost-effectiveness of nutrient-dense ONS containing high protein and β-hydroxy-β-methylbutyrate (β-HMB) compared with a placebo [15]. The outcomes included healthcare costs, measured as the product of resource use and per-unit cost, as well as quality-adjusted life-years (QALYs) at the 90-day time horizon, life-years (LYs) saved at the lifetime time horizon, and the incremental cost-effectiveness ratio (ICER). The results suggested that the tested

ONS cost-effectively extends the lives of malnourished hospitalized patients, as highlighted by a systematic review [16].

With respect to nursing homes, a randomized controlled trial in five nursing home facilities revealed that ONS were cost effective in increasing caloric intake but did not have a significant effect on body weight, despite positive trends [17].

Nonetheless, the effectiveness and cost-effectiveness of ONS with or without other dietary interventions for older people who are malnourished or at risk of malnutrition and frailty are still unclear [18]. Thus, there is a need for further studies reporting short-term and long-term health outcomes, determinants, and participant characteristics.

Finally, as far as Italy is concerned, a recent paper revealed that Foods for Special Medical Purposes (FSMPs) may be unaffordable for most older adults, given the average retirement income in Italy [19]. The effective use of FSMPs would likely require local prescription systems for hospital patients. This would shift the expense burden to the National Healthcare System [20, 21].

The Impact of Malnutrition on Healthcare Costs

Malnutrition is a major healthcare concern, with well-documented negative effects on clinical outcomes and overall quality of life [22]. It is associated with higher health care costs in both institutionalized or community-dwelling older adults [23]. In Italy, a systematic review of 67 studies examining the prevalence of DRM in Italian populations revealed an average prevalence of approximately 50% among hospitalized adults and 30% among hospitalized children [22]. Higher rates were observed in long-term care facility residents. Economic modeling suggested that the annual healthcare costs attributable to DRM in Italy exceed €10 billion in the base case scenario, with the most conservative estimates surpassing €2.5 billion. There are several reasons for the increase in hospital costs for malnourished patients. We can identify the following as the main ones [16, 24]:

(a) Increased hospital costs for longer hospital stays, increased risk of developing comorbidities, and increased risk of rehospitalization;
(b) Increased use of drugs and therapies;
(c) Increased costs for long-term care (home or residential care).

Disease-related malnutrition is known to have significant clinical and economic consequences. The systematic review conducted by Mitchell & Porter [24] aimed to establish the cost-effectiveness of identifying and treating malnutrition in the hospital setting. However, the authors reported that the number of economic evaluations for each type of intervention was limited; therefore, it was not possible to reach definite conclusions regarding the cost-effectiveness of interventions. Further studies are needed to pursue this direction.

However, as presented in the previous chapter, positive evidence has been reported on the effectiveness and cost-effectiveness of interventions based on ONS for tackling malnutrition in the hospital setting [15, 16].

Societal Costs of Malnutrition

Malnutrition also entails some indirect costs, including social costs. The worsening of health conditions imposes a greater burden on family members and caregivers, with effects on their quality of life, income and well-being.

The invisible indirect costs include premature mortality and productivity losses and equal 68% of the total costs, thus far exceeding the visible costs incurred by health and social care systems, equaling 32% [25]. Without interventions, the economic and societal consequences of overweight and obesity will be significant for countries across all regions. Considering projected population growth and shifts in age demographics, along with the continuation of current obesity trends, the global economic costs of overweight and obesity are expected to increase from just under US$ 2 trillion in 2020 to over US$ 3 trillion by 2030 and to exceed US$ 18 trillion by 2060 (at 2019 prices). Family caregivers play a vital role in maintaining the quality of life of older adults, especially those with cognitive impairment. For instance, a study by Tana et al. [26] revealed that impaired daily activities significantly increase caregiver burden, with poor nutrition in older outpatients being a stronger independent factor than cognitive or physical disability. Consequently, repercussions for the lives of families with subjects affected by malnutrition are important and need to be addressed. Implementing proper policies in this direction is fundamental. Below, we present several advanced policies addressing different targets.

Policies to Tackle Malnutrition

Achieving healthy, sustainable, and equitable diets is one of the challenges for twenty-first-century food systems. Here, we present some policies that have been highlighted in extant studies to promote a healthy diet and to address childhood malnutrition and older adults' malnutrition.

Policies for a Healthy Diet

To support global efforts, the WHO has updated methodologies for monitoring the affordability of healthy diets [27], enabling country-specific assessments critical to policy development aligned with the Sustainable Development Goals (SDGs) for 2030 [28].

In the European context, the WHO Regional Office for Europe developed the European Food and Nutrition Action Plan 2015–2020 [29] with the aim to combat the growing burden of diet-related NCDs, obesity, and malnutrition across the region. Emphasizing a whole-of-government, health-in-all-policies approach, the plan focused on strengthening food system governance and improving population nutrition. Moreover, the European Commission's publication titled *Review of Scientific Evidence and Policies on Nutrition and Physical Activity* [30] provided a comprehensive analysis of the factors influencing diet and physical activity across Europe, with a focus on addressing health inequalities. Released in 2019 by the Directorate-General for Health and Food Safety (DG SANTE) in collaboration with ICF Consulting Services, the report aimed to support policymakers in developing effective strategies to promote healthier lifestyles, particularly among vulnerable populations. The report analyzed policy interventions at different levels (systems/policies, retail settings, childcare and preschool settings, school and university settings, worksite settings and healthcare settings), resulting in several recommendations. In brief, those included the taxation of unhealthy products, e.g., sugar-sweetened beverages; high-fat/salt/sugar (HFSS) foods; subsidies and price reductions to make fruits, vegetables, and whole grains more affordable and accessible; limiting advertising of unhealthy foods to children and adolescents, especially through digital media and television; and the adoption of clear front-of-pack labeling (FoPL) to help consumers make healthier choices and mandatory nutritional guidelines for food served in school cafeteria, government facilities and public catering contracts. The population education should be consistent with the National Guidelines for Healthy Eating.

The WHO and other organizations have recommended taxes on sugar-sweetened beverages (SSBs) as part of a comprehensive set of policy measures to prevent diet-related NCDs. A recent study [31] aimed to evaluate the design, implementation, and impact of health-related taxes across EU member states, with a focus on products HFSS, including nonalcoholic beverages. The findings showed that such taxes reduce consumption and generate public revenue, some of which support health initiatives.

However, the study called for greater EU-wide coordination to avoid market fragmentation. Despite WHO recommendations, by 2020, only 10 of 53 European countries had implemented national taxes on SSBs [31].

Despite the promotion of FoPL at the European level [32] and the development of several guidelines on this topic [33], no harmonized system is currently in place across all European countries. The European Action Plan for the Prevention and Control of Noncommunicable Diseases 2016–2025 [34] provides policies for workplaces by highlighting the importance of safeguarding workers' well-being, including the management of psychological and work-related stress. In parallel, the European Network for Workplace Health Promotion (ENWHP), a European non-governmental organization with over 20 years of experience, has played a key role in developing innovative strategies to address workplace health challenges. Despite these efforts, there remains a need for uniform guidelines to support the effective implementation and monitoring of workplace health promotion programs across regions.

Importantly, some relevant policies to promote a healthy diet have been advanced in Italy. The Guidelines for a Healthy Diet, first introduced in 1986 by the National Institute for Nutrition and the Dietary Reference Values (DRVs), referred to in Italy as LARN and published by the Italian Society of Human Nutrition (Società Italiana di Nutrizione Umana - SINU) in 2024 [35], constitute foundational tools for national food policy in Italy. DRVs offer quantitative benchmarks for nutrient and energy intake aimed at meeting physiological requirements and preventing adverse health outcomes. Furthermore, guidelines providing food- and diet-based recommendations, informed by the Mediterranean dietary model [36], have been recently released to support the practical implementation of these nutritional targets within culturally appropriate eating patterns. Of note, updates to the Guidelines are periodically undertaken to reflect revisions to the DRVs [37].

In 2015, the city of Milan introduced the Milan Urban Food Policy Pact (MUFPP),[1] an international agreement among cities worldwide committed to developing sustainable food systems that are inclusive, resilient, safe, and diverse. The pact aims to ensure access to healthy and affordable food for all within a human rights framework while minimizing waste, conserving biodiversity, and addressing the impacts of climate change.

However, a critical analysis of the food policies implemented by cities participating in the pact reveals that signing the MUFPP does not always translate into concrete actions, particularly in the area of obesity prevention. Loperfido et al. [38] argued that the current food policy approach is flawed in its monitoring indicators, which are insufficient for evaluating, quantifying, and tracking the impact of the strategy both in the short and long term. Additionally, while food policies primarily target primary schools, which focus on food education, physical activity, and the promotion of healthy and sustainable diets, they need to be better integrated into regional and national frameworks to increase their effectiveness.

Another key step forward in Italy's public health efforts is represented by the "Guadagnare Salute" program[2]which has been promoted by the National Institute of Health as part of the European strategy for the prevention and control of chronic diseases. The main goal of the program is to prevent chronic and degenerative diseases by encouraging healthier behaviors and reducing risk factors, with a focus on promoting healthy lifestyles and early screening. Furthermore, national monitoring systems for adults and children have been established to collect evidence on the prevalence of unhealthy lifestyles and diseases. These include (i) "Keep an Eye on Health" (OKkio alla Salute, 2023[3]), coordinated by the National Institute of Health, in collaboration with the Ministry of Education, University and Research; (ii) the

[1] https://www.milanurbanfoodpolicypact.org/

[2] https://www.epicentro.iss.it/guadagnare-salute/programma/contesto

[3] https://www.epicentro.iss.it/okkioallasalute/

Health Progress in Italian Local Authorities system (Sorveglianza PASSI[4] -Progessi delle Aziende Sanitarie per la Salute in Italia), which provides information about risk factors (including obesity), people's perceptions of health, and the delivery of health services to people aged 18 to 69; (iii) the "Silver Steps" system (PASSI d'Argento[5]), which is a surveillance system on the health and quality of life of people over 64; and iv) the large prospective cohort follow-up study "Heart" project (Progetto CUORE[6]) for monitoring cardiovascular risk factors and high-risk conditions in the Italian population through standardized physical examinations.

Finally, on May 7, 2025, the Chamber of Deputies approved Bill 741, and Italy became the first country to formally recognize obesity as a chronic disease, marking a significant step forward in public health policy. This recognition led to the inclusion of obesity in the Essential Levels of Assistance (LEA) within the Italian healthcare system, potentially improving access to care, treatment, and long-term management.

Concerning policies favoring the adoption of sustainable diets in Italy, a review of international, European, and national policy documents on sustainable and healthy diets has been conducted [39]. This review highlights that at the national level, Italy promotes several strategies, including the Guidelines for Healthy Eating (CREA), the Mediterranean Diet [36], and the National Guidelines for Food Education in Schools. Additionally, the National Strategy for Sustainable Development (SNSvS), aligned with the UN 2030 Agenda, supports the development of local, resilient, and low-emission agri-food systems. Despite these efforts, the review highlights the lack of an integrated national strategy that holistically combines nutrition, sustainability, economy, and culture. Current Italian food policies remain fragmented across sectors and are primarily based on education and awareness, with limited regulatory measures, such as those related to advertising, labeling, or fiscal policy [39].

Policies for Childhood Malnutrition

At the global level, some guidelines have been advanced to address childhood obesity. Here, we focus on those offered by UNICEF in the Prevention of Overweight and Obesity in Children and Adolescents (2019) and those identified by the World Health Organization [40]. Moreover, since 1981, the FAO/WHO has been publishing a document concerning the International Code of Marketing of Breast milk Substitutes & Guidelines on Nutrition, whose main objective is to guide through concrete actions toward healthy early-life nutrition. The main focus is the promotion of breastfeeding, the introduction of safe and appropriate complementary

[4] https://www.epicentro.iss.it/passi/

[5] https://www.epicentro.iss.it/passi-argento/

[6] https://www.cuore.iss.it/

feeding, and the restriction of marketing unhealthy food and drink products to children, key goals for ensuring an adequate diet during the first 1000 days of life. Nevertheless, today only a limited number of countries fully comply with these recommendations [41].

In parallel, the WHO [40] outlined six priority areas for addressing childhood obesity: (1) promoting healthy food intake; (2) encouraging physical activity; (3) preconception and pregnancy care; (4) early childhood diet and activity; (5) school-based interventions; and (6) weight management. A coordinated and integrated approach across these domains is essential to ensuring access to nutritious food and enabling environments that support healthier lifestyles.

At the European level, the European Union has made an Action Plan on Childhood Obesity 2014–2020 [42], and the European Economic and Social Committee International Partnerships provided policy recommendations emphasizing the following: (i) addressing socioeconomic inequalities; (ii) strengthening regulations on food marketing to children; (iii) promoting healthy urban and school environments; (iv) enhancing digital and audiovisual regulations; and (v) encouraging family and community involvement.

Furthermore, the WHO European Childhood Obesity Surveillance Initiative (COSI[7]) monitors trends in overweight and obesity among primary school-aged children over a 15-year period, providing standardized data to track progress in addressing the onset of childhood obesity.

Under the EU4Health program (2021–2027), several initiatives were funded to contribute to a healthier Europe (e.g., funding national authorities, health organizations and other bodies through grants and public procurement). For example, the Scheme Schools4Health project aims to foster a healthy school environment by transforming schools into health-promoting schools (HPSs) to address childhood obesity. Moreover, all 28 EU countries currently have national school-feeding policies, with half setting mandatory standards and the other half providing voluntary guidelines with variety in their application.

With reference to policies for the marketing of foods and nonalcoholic beverages to children, an evaluation of these recommendations in the European Region in 2018 [43] identified numerous gaps and emphasized the need for more comprehensive policies and greater support for Member States to implement them effectively. Therefore, the WHO Regional Office for Europe developed comprehensive protocols and templates. These resources provide standardized methods for assessing exposure and impact across various media platforms.

In line with the European perspective for the general population, the European Academy of Pediatrics and the European Childhood Obesity Group, in 2021, advocated for the immediate implementation of a mandatory, standardized, and interpretative FoPL nutrition labeling system across the European Union to increase consumer awareness of the nutritional quality of packaged foods. Therefore, these

---

[7] https://schools4health.eu/

groups urged EU legislators to introduce a mandatory, uniform, and interpretative FoPL nutrition labeling system [44].

As far as Italy is concerned, the National Center for Disease Prevention and Health Promotion of the Italian National Institute of Health participates in the Joint Action—Health4EUkids initiative.[8] In particular, its objectives include (i) increasing commitment to tackling obesity, (ii) assessing the integration of health promotion in national agendas, (iii) advancing a health-in-all-policies approach, iv) identifying enablers and barriers to policy implementation, v) engaging policymakers in dialog, and vi) supporting the incorporation of joint action outcomes into national and EU-level strategies.

In line with this, since 2016, Milan's school canteens have implemented the "**Fruit for Mid-Morning Snack**" project,[9] which aims to completely replace the snacks typically brought from home with fresh fruit, served during the mid-morning break instead of after lunch. In the 2022–2023 school year, the initiative involved 54 primary schools and one secondary school, reaching a total of 14,613 students. In addition to promoting healthier eating habits, the project also led to a 17% reduction in food waste.[10]

References

1. Lenoir-Wijnkoop I, Dapoigny M, Dubois D, van Ganse E, Gutiérrez-Ibarluzea I, Hutton J et al (2011) Nutrition economics–characterizing the economic and health impact of nutrition. Br J Nutr 105(1):157–166
2. Martinez R, Lloyd-Sherlock P, Soliz P, Ebrahim S, Vega E, Ordunez P, McKee M (2020) Trends in premature avertable mortality from noncommunicable diseases for 195 countries and territories, 1990–2017: a population-based study. Lancet Glob Health 8(4):e511–e523
3. Jakovljevic M, Getzen TE (2016) Growth of global health spending share in low and middle income countries. Front Pharmacol 7:21
4. FAO, IFAD, UNICEF, WFP, WHO (2019) The state of food security and nutrition in the world 2019. In: Safeguarding against economic slowdowns and downturns. FAO, Rome
5. FAO (2023) Cost and affordability of a healthy diet. https://openknowledge.fao.org/server/api/core/bitstreams/f1ee0c49-04e7-43df-9b83-6820f4f37ca9/content/state-food-security-and-nutrition-2023/cost-affordability-healthy-diet.html
6. FAO (2024) Food security and nutrition in the world. https://openknowledge.fao.org/items/ebe19244-9611-443c-a2a6-25cec697b361
7. World Health Organization (2021) Draft recommendations for the prevention and management of obesity over the life course, including potential targets. https://cdn.who.int/media/docs/default-source/obesity/who-discussion-paper-on-obesity—final190821.pdf?sfvrsn=4cd6710a_1&download=true
8. Ling J, Chen S, Zahry NR, Kao TSA (2023) Economic burden of childhood overweight and obesity: a systematic review and meta-analysis. Obes Rev 24(2):e13535

[8] https://www.epicentro.iss.it/en/obesity/joint-action-health4EUkids

[9] https://www.milanoristorazione.it/progetti-educativi/frutta-a-meta-mattina

[10] https://www.comune.milano.it/aree-tematiche/food_policy/refezione-scolastica-e-educazione-alimentare

9. Pelone F, Specchia ML, Veneziano MA et al (2012) Economic impact of childhood obesity on health systems: a systematic review. Obes Rev 13(5):431–440
10. Llewellyn A, Simmonds M, Owen CG, Woolacott N (2016) Childhood obesity as a predictor of morbidity in adulthood: a systematic review and meta-analysis. Obes Rev 17(1):56–67
11. Cohen AK, Rai M, Rehkopf DH, Abrams B (2013) Educational attainment and obesity: a systematic review. Obes Rev 14(12):989–1005
12. Segal AB, Huerta MC, Aurino E, Sassi F (2021) The impact of childhood obesity on human capital in high-income countries: a systematic review. Obes Rev 22(1):e13104
13. Dent E, Wright OR, Woo J, Hoogendijk EO (2023) Malnutrition in older adults. Lancet 401(10380):951–966
14. Volkert D, Beck AM, Cederholm T, Cruz-Jentoft A, Hooper L, Kiesswetter E, Maggio M, Raynaud-Simon A, Sieber C, Sobotka L, van Asselt D, Wirth R, Bischoff SC (2022) ESPEN practical guideline: clinical nutrition and hydration in geriatrics. Clin Nutr 41(4):958–989
15. Zhong Y, Cohen JT, Goates S, Luo M, Nelson J, Neumann PJ (2017) The cost-effectiveness of oral nutrition supplementation for malnourished older hospital patients. Appl Health Econ Health Policy 15:75–83
16. Elia M, Normand C, Laviano A, Norman K (2016) A systematic review of the cost and cost effectiveness of using standard oral nutritional supplements in community and care home settings. Clin Nutr 35(1):125–137
17. Simmons SF, Keeler E, An R, Liu X, Shotwell MS, Kuertz BRD, Silver HJ, Schnelle JF (2015) Cost-effectiveness of nutrition intervention in long-term care. J Am Geriatr Soc 63(11):2308–2316
18. Thomson KH, Rice S, Arisa O, Johnson E, Tanner L, Marshall C, Sotire T, Richmond C, O'Keefe H, Mohammed W, Raffle A, Hanratty B, McEvoy CT, Craig D, Ramsay SE (2022) Effectiveness and cost-effectiveness of oral nutritional supplements in frail older people who are malnourished or at risk of malnutrition: a systematic review and meta-analysis. Lancet Healthy Longev 3(10):e654–e666
19. Madini N, Vincenti A, Beretta A, Santero S, Viroli G, Cena H (2024) Addressing inflammaging and disease-related malnutrition: adequacy of oral nutritional supplements in clinical care. Nutrients 16(23):4141
20. Cavazza M, Banks H, Muscaritoli M, Rondanelli M, Zandonà E, Jommi C (2020) Patient access to oral nutritional supplements: which policies count? Nutrition 69:110560
21. Perugini M, Johnson TJ, Beume TM, Dong OM, Guerino J, Hu H, Kerr K, Kindilien S, Nuijten M, Ofili U, Taylor MT, Wong A, Freijer K (2022) Are we ready for a new approach to comparing coverage and reimbursement policies for medical nutrition in key markets: an ISPOR special interest group report. Value Health 25(5):677–684
22. Pradelli L, Zaniolo O, Sanfilippo A, Lezo A, Riso S, Zanetti M (2023) Prevalence and economic cost of malnutrition in Italy: a systematic review and metanalysis from the Italian Society of Artificial Nutrition and Metabolism (SINPE). Nutrition 108:111943
23. Abizanda P, Sinclair A, Barcons N, Lizán L, Rodríguez-Mañas L (2016) Costs of malnutrition in institutionalized and community-dwelling older adults: a systematic review. J Am Med Dir Assoc 17(1):17–23
24. Mitchell H, Porter J (2016) The cost-effectiveness of identifying and treating malnutrition in hospitals: a systematic review. J Hum Nutr Diet 29(2):156–164
25. Okunogbe A, Nugent R, Spencer G, Powis J, Ralston J, Wilding J (2022) Economic impacts of overweight and obesity: current and future estimates for 161 countries. BMJ Glob Health 7(9):e009773
26. Tana C, Lauretani F, Ticinesi A, Gionti L, Nouvenne A, Prati B et al (2019) Impact of nutritional status on caregiver burden of elderly outpatients. A cross-sectional study. Nutrients 11(2):281
27. Bai Y, Herforth A, Cafiero C, Conti V, Rissanen MO, Masters WA, Rosero Moncayo J (2024) Methods for monitoring the affordability of a healthy diet. In: FAO statistics working paper series, no. 24–44. FAO, Rome

28. FAO, IFAD, UNICEF, WFP, WHO (2024) The State of Food Security and Nutrition in the World 2024—financing to end hunger, food insecurity and malnutrition in all its forms, Rome
29. World Health Organization (2015) European food and nutrition action plan 2015–2020
30. European Commission (2018) Directorate-General for Health and Food Safety and ICF Consulting Services, Review of scientific evidence and policies on nutrition and physical activity—A comprehensive review of the scientific evidence about the source of calories consumed and types and frequency of physical activity among Europeans, Publications Office of the European Union
31. European Union (2025) Health taxes from an EU perspective. https://op.europa.eu/en/publication-detail/-/publication/93844ea8-3133-11f0-8a44-01aa75ed71a1/language-en
32. European Commission (2020) Report from the Commission to the European Parliament and the Council regarding the use of additional forms of expression and presentation of the nutrition. https://food.ec.europa.eu/document/download/310ad7a1-aa8b-49f6-babf-55ec572f0f75_en?filename=labellingnutrition_fop-report-2020-207_en.pdf
33. Storcksdieck GBS, Marandola G, Ciriolo E, Van BR, Wollgast J (2020) Front-of-pack nutrition labelling schemes: a comprehensive review
34. WHO (2016) Action plan for the prevention and control of noncommunicable diseases in the WHO European region. In: Action plan for the prevention and control of noncommunicable diseases in the WHO European Region
35. SINU. LARN. (2024) Livelli di Assunzione di Riferimento di Nutrienti ed energia per la popolazione italiana. V revision. BioMedia editor
36. Istituto Superiore di Sanità (2025) Linea guida—La dieta Mediterranea. Available online at: https://www.iss.it/documents/20126/10024898/LG+C0035-SINPE+et+al.pdf/da49883e-c0ce-ba5f-0627-91a02b0f2bfd?t=1744374219082. Accessed 9 Jun 2025
37. EFSA (2023). Dietary reference values (DRVs), https://www.efsa.europa.eu/en/topics/topic/dietary-reference-values
38. Loperfido F, Vincenti A, Basilico S, Conti MV, Orsini F, Franchi C et al (2025) Rethinking food policy framework to tackle childhood obesity in the Italian context. Nutr Rev 83(2):e1–e3
39. Mazzocchi G (2024) Policies for healthy and sustainable diets: a review of international, European, and national documentation. CREA Council for Agricultural Research. https://www.crea.gov.it/documents/68457/0/2024_Policies+for+HSD_OnFoods.pdf/39d1c771-3549-c386-6834-65b1a5b1cd16?t=1717498668433%3E
40. WHO (2019) Essential nutrition actions: mainstreaming nutrition through the life-course. Available at https://www.who.int/publications/i/item/9789241515856
41. Becker GE, Zambrano P, Ching C, Cashin J, Burns A, Policarpo E, Datu-Sanguyo J, Mathisen R (2022) Global evidence of persistent violations of the international code of marketing of breast-milk substitutes: a systematic scoping review. Matern Child Nutr 18(Suppl 3):e13335
42. EU Action Plan on Childhood Obesity 2014-2020. https://health.ec.europa.eu/document/download/7c75c967-69c0-4cc2-8906-bc8b59f58a79_en?filename=childhoodobesity_actionplan_2014_2020_en.pdf
43. World Health Organization. Regional Office for Europe (2018) Evaluating implementation of the WHO Set of recommendations on the marketing of foods and non-alcoholic beverages to children: progress, challenges and guidance for next steps in the WHO European region. World Health Organization. Regional Office for Europe
44. Dereń K, Dembiński Ł, Wyszyńska J, Mazur A, Weghuber D, Łuszczki E et al (2021) Front-of-pack nutrition labeling: a position statement of the European academy of pediatrics and the European childhood obesity group. Ann Nutr Metab 77(1):23–28

Open Access This chapter is licensed under the terms of the Creative Commons Attribution-NonCommercial-NoDerivatives 4.0 International License (http://creativecommons.org/licenses/by-nc-nd/4.0/), which permits any noncommercial use, sharing, distribution and reproduction in any medium or format, as long as you give appropriate credit to the original author(s) and the source, provide a link to the Creative Commons license and indicate if you modified the licensed material. You do not have permission under this license to share adapted material derived from this chapter or parts of it.

The images or other third party material in this chapter are included in the chapter's Creative Commons license, unless indicated otherwise in a credit line to the material. If material is not included in the chapter's Creative Commons license and your intended use is not permitted by statutory regulation or exceeds the permitted use, you will need to obtain permission directly from the copyright holder.

Chapter 4
Healthy Eating Strategies in the Workplace: An Organizational Perspective

Beatrice Re and Pietro Previtali

Corporate Wellness Programs and Employees' Well-being

CW programs are defined as a "combination of educational, organizational, and environmental activities designed to motivate and promote healthy lifestyles among employees and their families" ([1], p. 201).

The implementation of CW programs aligns with the principles of organizational support theory (OST), i.e., by investing in employee health and well-being, organizations demonstrate a commitment that extends beyond the basic employment contract. This act of care and support can significantly enhance Perceived Organizational Support (POS), activating the reciprocity mechanism described in the OST. Employees who perceive that their organization is investing in their well-being are more likely to reciprocate with increased commitment and performance [2].

Corporate wellness (CW) programs have become a key strategy for organizations aiming to boost perceived organizational support (POS) and improve employee outcomes. The substantial financial commitment devoted to CW programs highlights the increasing recognition of employee well-being as a key factor for organizational success. Indeed, the CW industry is now a strong sector valued at USD 65 billion [3]. The industry's expansion is mainly due to rising awareness of employee health and well-being, along with the growing costs of healthcare. In response, organizations are increasingly investing in wellness initiatives to boost productivity, reduce absenteeism, and enhance overall employee satisfaction, thereby driving a notable increase in market demand.

The market landscape is dominated by actors primarily concentrated in North America and Europe. The United States and Canada hold leading positions, supported by widespread employer adoption and well-established healthcare benefit

B. Re (✉) · P. Previtali
Department of Economics and Management, University of Pavia, Pavia, Italy
e-mail: beatrice.re@unipv.it; pietro.previtali@unipv.it

© The Author(s) 2026

B. Re, P. Previtali (eds.), *Tackling Malnutrition in a Complex World*,
SpringerBriefs in Business, https://doi.org/10.1007/978-3-032-26438-1_4

systems. Similarly, Europe—particularly the United Kingdom and Germany—represents a significant market, sustained by high levels of employer-sponsored wellness programs and a growing emphasis on mental health and digital wellness solutions.

The positive impact of CW programs on both employees and organizations is well-documented. Gubler et al. [4] demonstrated that CW programs, along with health improvements, can increase employee productivity through two main factors: job motivation and capability. Their research revealed that employees with preexisting health issues who saw improvements through wellness programs experienced a 10% increase in productivity. This finding supports the concept of reciprocity, as these employees, who feel grateful, are more likely to work harder in return [5, 6].

The connection between CW programs and OST is further reinforced by Guest's [7] model, which explores the factors that contribute to employee well-being. This model highlights elements such as "investing in employees," "providing engaging work," "positive social and physical environment," "voice," and "organizational support" as key contributors. CW programs serve as a concrete expression of these elements, positively affecting both employee well-being and the overall outcomes in the model, particularly in terms of "individual and organizational performance."

Additionally, Grawitch et al. [8] developed the "PATH model framework" to synthesize previous research on workplace health, categorizing organizational practices into "healthy workplace practices," which closely align with CW programs. Among these practices is "health and safety," which affects employee well-being (including physical health, mental health, stress, motivation, job satisfaction, and morale) as well as organizational improvements (e.g., performance/productivity, absenteeism, and turnover).

However, there are potential drawbacks and paradoxes associated with CW programs. Critics argue that these programs can intensify the link between wellness initiatives and economic interests, overemphasizing economic performance and increasing employer control over employees [9–11]. Furthermore, as noted by Kavaratzis et al. [12], paradoxes may arise when well-being is balanced with other workplace outcomes, such as job performance and satisfaction. These critiques highlight the need for CW programs to prioritize employee well-being rather than focusing solely on organizational goals.

Among the initiatives within CW programs, diet and nutrition are critical not only for physical well-being but also for mood and mental health [13]. However, many employees struggle to maintain healthy diets because of time constraints, work stress, and the availability of unhealthy food options. This is problematic, as poor nutrition can lead to decreased productivity, increased turnover, early retirement, and increased healthcare costs [14]. Orfanos et al. [15] reported that employees who eat only at work, without dining at restaurants, do not typically see the increase in energy intake that eating out generally causes [16]. Therefore, addressing employees' eating habits within the workplace can have a meaningful impact.

In Italy, where food is central to daily life and eating well is an integral part of the culture, more than 70% of workers take a lunch break every day. Forty-one percent of them eat lunch at their workplace (e.g., homemade meals, canteens, or vending machines), whereas 59% eat outside (e.g., in nearby restaurants and cafes) [17]. While the demand for healthy food is growing among European employees [18], Italian workers are particularly discerning, often seeking high-quality, organic, and sustainable food options, with an emphasis on vegetables (ibid). Investing in healthy eating initiatives at the corporate level can help bridge socioeconomic gaps among employees, which may influence food choices [19]. Companies in Italy, for example, have introduced the "distributed canteen" concept, allowing employees to access local partner restaurants through an app, which also provides nutritional information [20].

A growing number of organizations are incorporating diet and nutrition initiatives into their CW programs. Such initiatives often involve providing healthier food and beverage options in workplace cafeterias [21] to reduce the risks of malnutrition and diet-related illnesses, including diabetes [22]. According to Mulaney et al. [23], participation in diet and nutrition programs ranks highest among CW initiatives, indicating that employees particularly appreciate these opportunities. From a human resources (HR) standpoint, ensuring access to nutritious meals-even if offered at a cost-can enhance employees' quality of life and work while also benefiting their families [22]. Moreover, prioritizing employee well-being helps cultivate a positive organizational climate, which in turn fosters higher levels of engagement and commitment [24].

Although the health and medical dimensions of diet and nutrition have been extensively studied (e.g., [25–28]), their organizational and managerial implications remain underexplored [4]. Research in organizational studies typically adopts a broader focus on employee well-being-encompassing aspects such as happiness, health, and social relationships-and how these influence performance and retention [29, 30]. Drewnowski [26] proposed a model that connects health determinants, including diet and nutrition, with workplace productivity, alongside mental health and musculoskeletal conditions. Building on this, Gubler et al. [4] examined nutrition and diet initiatives as integral elements of CW programs, highlighting their potential to boost productivity and organizational profitability. Nonetheless, further research is needed to better understand how such programs shape employee reciprocity [4] and their broader relationship with individual well-being and organizational outcomes [7].

Nutrition and Mental Health in the Workplace

The relationship between nutrition and mental health is multifaceted. The emerging field of psycho-nutrition has garnered increasing scholarly attention, driven by accumulating evidence that underscores the positive influence of healthy dietary patterns, reduced caloric intake, and fasting practices on mental well-being [31]. In particular, caloric restriction and fasting regimens have been identified as highly effective non-pharmacological interventions for enhancing both cardiovascular and metabolic health outcomes [32].

Emotions exert a significant impact on eating behavior. Emotional regulation, which encompasses both automatic and deliberate processes, plays a central role in shaping the onset, persistence, intensity, and duration of affective states [33]. Individuals employ a variety of regulation strategies to cope with emotions, one of which is food consumption. Unhealthy eating behaviors may function as a rapid, emotion-driven coping mechanism, motivated by the expectation of immediate mood enhancement [34]. Although often regarded as irrational, emotional eating is a widely documented phenomenon across diverse populations. Substantial empirical evidence highlights the close association between emotions and eating patterns, with pronounced effects observed in individuals with eating disorders [35, 36]. For instance, individuals with obesity have been found to increase their food intake in response to negative affective states [37]. Similarly, "restrained eaters"—those who adhere to chronic dieting—are particularly vulnerable to overeating when confronted with adverse emotional circumstances (e.g., [38, 39]). Nevertheless, emotional eating is not restricted to clinical populations, as it has also been observed among otherwise healthy individuals [40].

According to emotional eating theory (e.g., [41, 42]), negative affective states heighten the motivation to eat, with food intake subsequently attenuating the intensity of such emotions. Stress is one of those states that has been consistently identified as a salient trigger of emotional eating, with workplace stress exerting especially detrimental effects on dietary behavior [43]. Empirical studies confirmed that occupational stress is positively associated with unhealthy eating practices [44, 45]. Sonnentag et al. [45] found that employees experiencing high-stress workdays requiring substantial self-regulation were more likely to consume sweets to regulate their emotions. Similarly, boredom and stress within the workplace context have been linked to elevated caloric intake during working hours [44].

In their review, Clohessy et al. [46] identified a range of factors shaping employees' dietary behaviors in office-based environments, including job-related demands (e.g., high-pressure settings, employer-provided meals, and stress), features of the workplace food environment (e.g., limited availability of healthy options, easy access to energy-dense foods, and lack of preparation facilities), as well as social influences. Facilitators of healthy eating included colleague support and environmental modifications such as providing free fruit, expanding healthy meal options, and reducing the availability of unhealthy alternatives. These factors frequently interact; for instance, workplace boredom combined with easy access to confectionery, or extended work hours coupled with social norms, can shape employees' dietary decisions.

Given that employees spend a considerable proportion of their time at work and consume approximately one-third of their daily caloric intake during working hours (ibid), organizations can significantly influence employee health and well-being through the promotion of diet-related interventions. A variety of initiatives have been experienced by firms in this direction, some of which are outlined below.

Promoting Healthy Eating in Organizations: Case Studies

We here introduce some case studies of organizations that experimented with different healthy eating initiatives. Table 4.1 synthesizes our case studies and the data we collected between 2024 and 2025, by highlighting the interviewees' roles and interview details.

Case Study CIRFOOD

With more than 50 years of experience, CIRFOOD[1] is among the largest Italian companies in collective catering, commercial catering, and welfare services. The firm operates in 18 Italian regions and 75 provinces, as well as internationally in Belgium and the Netherlands. While particularly specialized in school catering—from nurseries to middle schools—it also serves universities, companies, healthcare organizations, hospitals, and retirement homes. In addition, CIRFOOD manages commercial catering chains (public establishments) and self-service restaurants designed for employees' lunch breaks.

Because it primarily collaborates with public institutions, the company must comply with national guidelines (e.g., *lex specialis*) when designing menus. As highlighted by the head of the design and service development office, such regulations may constrain menu composition and product choices. Nonetheless, the interviewee emphasized that collective catering also plays an educational role in promoting proper nutrition:

Table 4.1 Comprehensive data collection. Source: Authors' own work

	Type of data	Interviewee's role	Details
Primary data	2 in-depth interviews	Communication Specialist, Quality and Production Systems Manager, Head of the Design Office and Development of service Offerings—Cirfood	September 2024—1.30 h
	3 in-depth interviews	Nutrition Office Manager—Elior-Hospes Hospitality Specialist—Galdi HR Manager—Baxi	February–March 2025—1.15 h
	1 in-depth interview	Communication Manager—PlanEat HR managers—BeSharp	May 2024—2 h
	1 in-depth interview	Director of Pellegrini's Academy, Pellegrini Group	October 2024–1 h
Secondary data	Companies' websites and web articles		/

[1] https://www.cirfood.com/

> *One of the roles of collective catering, therefore also of CIRFOOD, is to educate on correct nutrition, also by supporting contracting bodies in identifying food proposals that lead to correct nutrition.*

In line with this mission, CIRFOOD launched the project "Less is More: The Sustainable Value of Hospital Catering" in 2023, with the contribution of Professor Andrea Pezzana. This initiative targets hospital catering with the dual aim of improving patients' nutritional intake while reducing food waste. According to the quality and production systems manager, the project also pursues an important cultural change:

> *We aim to replace the classic composition of first and second side dishes plus bread and fruit with a nice single dish, which is a concept we struggle so much to communicate also in school catering. The single dish is our battle, because it fights against waste, because it nourishes in the right way, because it gives you the possibility of ensuring the intake of legumes and cereals and other foods which are otherwise more difficult to propose.*

Another flagship initiative is the "Nutrition and Taste Focus on Dysphagia" project, developed in collaboration with Mysurable, an academic spin-off of the University of Bologna specializing in malnutrition and muscle loss in the elderly. Initially implemented in retirement homes and later extended to hospitals in the Liguria region, this project aimed to improve the quality of life of patients with dysphagia while reducing healthcare costs. As the quality and production systems manager explained, the project's key challenge was ensuring menus that were both balanced and texturally consistent. By adopting specialized equipment capable of processing ingredients at very high speeds, CIRFOOD succeeded in creating a scalable model that ensures constant textures across a variety of recipes. According to the head of the design and development office, the project not only addressed nutritional needs but also enhanced the human dimension of care, making meals more enjoyable for patients and thereby tackling malnutrition more effectively.

Beyond healthcare, CIRFOOD is deeply engaged in school catering, with a strong commitment to combating childhood obesity. The company designs balanced menus supported by nutritional education initiatives targeted at different age groups-from infants to secondary school students. As the head of the design office explained:

> *We offer nutritional education activities aimed at various age groups... We also aim to directly educate children on what are the best choices, obviously encouraging them also to other types of nutrition. We also promote a plant-based diet, for example, raising awareness of the consumption of legumes.*

In 2024, CIRFOOD also partnered with IULM University and the University of Gastronomic Sciences of Pollenzo on an experimental project at the CIRFOOD District restaurant, exploring how external factors such as background music can enhance meal enjoyment and reduce food waste. Additionally, in collaboration with CREA, the firm has contributed to advancing sensorial and environmental sustainability by training dietitian nutritionists who promote nutritional education in schools.

CIRFOOD also designs customized menus for companies, tailoring caloric and nutritional content to the needs of white- and blue-collar employees. Compared with public catering, private contracts allow greater flexibility in promoting sustainability initiatives, such as offering more plant-based options.

> *Here, there is more possibility to propose some projects that also aim to raise awareness concerning nutritional and environmental sustainability. For instance, we may propose slightly more plant-based dishes... because in private deals, we have greater freedom of negotiation.*

Overall, CIRFOOD represents a virtuous example of a collective catering company that experiments with innovative solutions to address malnutrition while balancing regulatory requirements, customer needs, and sustainability goals.

Case Study Elior-Hospes

To explore the implementation of workplace healthy eating initiatives, we examined the case of Elior Hospes,[2] a leading actor in the collective catering sector. In 2024, Elior was recognized as Italy's foremost collective catering company in the Italy's Best Employers survey. To gain a deeper understanding of its practices, we conducted in-depth interviews with the manager of Elior's Nutrition Office and with representatives from two client firms. The latter were selected using a snowball sampling strategy [47], beginning with referrals provided by Elior's Nutrition Office manager. Data were collected between February and March 2025 through three Zoom interviews (see Table 4.2 for an overview of data collection).

The first interview, conducted with the Nutrition Office manager, investigated how Elior supports its client organizations in promoting healthy nutrition, with particular attention to menu planning, targeted initiatives, and communication strategies. The second and third interviews were held with the hospitality specialist at Galdi and the HR manager at Baxi, two of Elior's client firms. These discussions explored how partner companies implement healthy eating initiatives in collaboration with Elior, as well as the perceived benefits for employee well-being and organizational performance.

Table 4.2 Data collection Elior Hospes. (Source: Authors' own work)

	Type of data	Interviewee's role	Details
Primary data	In-depth interview	Nutrition Office Manager/Elior	February 2025/1 h
	In-depth interview	Hospitality Specialist/Galdi	March 2025/1 h
	In-depth interview	HR Manager/Baxi	March 2025/1.15 h
Secondary data	Companies' websites and web articles		/

[2] https://www.elior.it/

Collectively, the interviews shed light on how Elior and its partner firms frame healthy eating not only as a nutritional benefit but also as part of an organizational strategy to achieve a competitive advantage.

Elior provides catering services to a wide range of organizations-small, medium, and large-across sectors such as education, healthcare, leisure, and business & industry. The company places particular emphasis on promoting healthy eating in the workplace, as reflected in its website statement:

> *Eating well means feeling good, even at work. The lunch break is a crucial part of the day, often influencing its flow, and helps people stay focused, attentive, and productive. Increasingly, companies view catering services as part of corporate welfare.*

Currently, Elior employs approximately 8000 people and operates under several brands, including Hospes, which is active in Northeast Italy, particularly the Veneto region. Galdi,[3] located in Treviso (Veneto region), is a producer of packaging solutions for the food industry and employs 130 people. Known for its innovative culture, the company continuously improves efficiency through the adoption of Lean and Agile methodologies. In January 2022, in collaboration with Hospes, Galdi inaugurated the *Galdi Bistrot*—a catering facility designed to promote healthy and natural food for employees. According to the hospitality manager (company website):

> *Galdi Bistrot complements the corporate welfare initiatives developed at the Village. A healthy diet forms the foundation of both quality of life and work. We conceived the lunch break as a moment to care for ourselves, with genuine and balanced dishes—without compromising on taste and dish presentation. After all, we are Italian!*

Baxi,[4] with almost a century of experience in the heating and air conditioning sector, employs over 850 people and reported a turnover of €380 million in 2024. The company operates extensively in Italy and exports its products to more than 50 countries worldwide. Baxi increasingly views nutrition as integral to employee well-being, positioning the canteen as a strategic corporate asset. Its collaboration with Hospes began a decade ago, transforming the canteen from a basic dining facility into a social hub offering healthy and sustainable food options.

Following the interviews, data were analyzed using three dimensions of interest: (i) healthy eating initiatives, (ii) individual well-being, and (iii) organizational performance. The findings show that Elior implements several initiatives for its client firms, including:

- Nutritional desk: a service offering dietary advice to employees;
- Nutrition seminars: addressing both general themes and specific issues such as nutrition and sport, celiac disease, and eating disorders;
- Thematic weeks: focused on nutrition-related topics, supported by posters and informational materials.

[3] https://www.galdi.it/en/

[4] https://www.baxi.com

Table 4.3 Data analysis. (Source: Re & Previtali [48])

	Healthy eating initiatives	Individual well-being	Organizational performance
Galdi	Introduction of balanced and varied menus	Improve employee health, increase in concentration	Increase productivity, improve job satisfaction
	Food education seminars	Improve employee health	Increase productivity
	Meetings with a nutritionist	Improve employee health	Improve job satisfaction
	Transition from the canteen to the opening of a modern *bistrot* with a chef	Relaxation and leisure	Improve job satisfaction Reduce employees' turnover
Baxi	Introduction of balanced and varied menus	Improve employee health, increase in concentration	Reduce absenteeism, increase productivity
	Meetings with a nutritionist	Improve employee health	Improve job satisfaction

In addition, Elior communicates the sustainability dimension of nutrition by indicating the CO_2 impact of recipes on its menus. While menu satisfaction surveys are conducted, the company does not systematically monitor the broader effects of its initiatives on communication outcomes or behavioral changes (e.g., dietary habits).

The analysis of the interviews with Galdi and Baxi—as reported in the article by Re and Previtali [48] (Table 4.3 below)—highlight some healthy eating initiatives implemented in the workplace and their expected impact on employee well-being and organizational performance.

Our analysis shows that firms are experimenting with a range of healthy eating initiatives, such as introducing balanced and varied menus, organizing food education seminars and nutritionist-led sessions, and opening bistros run by professional chefs. These initiatives are designed to generate benefits both at the individual level-enhancing employee health, concentration, relaxation, and leisure-and at the organizational level-boosting job satisfaction, productivity, and reducing absenteeism and turnover. Most of these outcomes are consistent with those reported in the literature [4, 49].

A novel insight, however, concerns the role of "relaxation and leisure" as a dimension of individual well-being, particularly linked to the presence of a modern bistro. Unlike a conventional canteen, this space is conceived as an experiential environment, where employees can enjoy scenic views of the mountains and access a roof garden, transforming breaks into moments of mindfulness. Offering employees a place where they can seek individual well-being through relaxation and leisure during breaks may represent a key driver for organizations seeking to enhance job satisfaction and reduce staff turnover.

Another noteworthy finding lies in employees' increasing interest in the environmental implications of food consumption, which has prompted firms to act in this area. For instance, through Elior-Hospes, they provide plant-based alternatives to traditional meals and adopt food recovery services such as TooGoodToGo.[5] These practices align with Italian workers' pursuit of high-quality food and their growing

[5] https://www.toogoodtogo.com/it

concern for sustainability [18]. Nonetheless, several barriers remain: adopting healthier diets, especially those centered on balanced or plant-based dishes, often requires changing habits that are difficult to achieve in more traditionalist contexts. To address this barrier, Elior-Hospes supports firms in raising awareness about sustainability by displaying information on calories and the CO_2 emissions of each dish directly on menus.

Finally, our findings suggest that offering healthy eating initiatives either free of charge (Baxi) or at a minimal contribution (Galdi) can be an effective means of improving job satisfaction while simultaneously addressing socioeconomic inequalities among employees [19]. This represents a novel finding that deserves further investigation.

Case Study PlanEat and BeSharp

PlanEat.eco,[6] operated by Alimentiamoci Srl Benefit Corporation, is an innovative Milan-based start-up founded in 2019. Active in the food sector, the company focuses on developing, producing, and marketing eco-friendly products and services, while generating value for local communities. At the heart of PlanEat's operations is a platform that delivers local, low-impact, portioned food, enabling customers to easily plan meals and receive ingredients at home in compostable packaging.

Recently, PlanEat has expanded into the Business to Business (B2B) market by offering sustainable and healthy lunch solutions to corporate employees. This move makes the company a compelling example of how businesses can promote employee well-being through nutrition-focused initiatives.

The company's sales and communications manager describes the service as follows:

> *We've created a platform for companies that offers a healthy, innovative, waste-conscious lunch break, aiming to drive change that protects people, work, and the planet. It is a fully customizable, market-first service—flexible and tailor-made so that each partnership can be uniquely designed. We also provide impact reports showing each customer's environmental footprint.*

Among its latest innovations, PlanEat has introduced reusable containers that are collected, washed, and reused. Once they reach the end of their lifespan, the containers are melted down and remanufactured, effectively eliminating packaging waste. Another key innovation is a software platform connecting restaurants and food providers directly with company employees. Through it, employees can preorder meals from a broader menu, allowing providers to prepare exactly what is needed-cutting down food waste and ensuring more efficient and sustainable deliveries while concurrently guaranteeing fair working conditions for delivery riders.

[6] https://planeat.eco/

Moreover, ordering from PlanEat allows employees to choose from a variety of meals and accommodate specific food choices such as veganism, thus promoting inclusive food practices.

One of PlanEat's corporate clients is beSharp, an IT company founded in 2011 in Pavia (south of Milan). With approximately 60 employees, beSharp operates in the B2B tech sector and is dedicated to helping companies embrace cutting-edge technological models to achieve their most ambitious goals. As part of its employee welfare initiatives, beSharp has incorporated diet-focused services to encourage healthier eating habits. Employees can preorder meals through PlanEat's platform, schedule their meals weekly, and store them in compostable containers in a dedicated office fridge-offering a practical and sustainable workplace nutrition solution. PlanEat is part of Sharp's welfare services and contributes to better nutrition at work. By conducting interviews with PlanEat communication specialists and Sharp's HR managers, followed by a fuzzy set qualitative comparative analysis of 20 questionnaires completed by Sharp's employees, the study by Re et al., (2025) [50] revealed that a successful approach to achieving both employees' well-being and organizational performance involves maintaining a balanced diet, ensuring high service quality, and reducing the duration of lunch breaks, while minimizing lunch-related management stress scores low or is absent. Some empirical studies highlight that employees tend to view diet-related interventions positively, particularly because of perceived increases in productivity. These results contribute to organizational support theory on a theoretical level and provide valuable practical guidance for management.

Case Study: Pellegrini S.p.a. and FOOD-AUT Project

Pellegrini is an Italian company founded in 1965 by Ernesto Pellegrini, which has 10,000 employees in Italy and around the world and a turnover of 700 million euros. The company is involved in the management of services, with a significant impact on people's quality of life: catering, vending, food supplies, cleaning and integrated services, and welfare solutions. Since its establishment, Pellegrini has stood out for service on the basis of quality and the constant search for excellence through its Academy. Since 2018, Pellegrini has taken a path toward sustainability, which has led to the definition of materiality issues.

In 2023, the Pellegrini Group started a scientific research project concerning nutrition in autistic individuals.[7] The project was coordinated by **Pellegrini Academy**, a center for research, development, and training where experts, educators, and researchers collaborate to generate innovative and practical proposals, all while upholding the Group's core values, particularly its commitment to social

[7] https://sostenibilita.gruppopellegrini.it/food-aut-per-una-migliore-alimentazione/

responsibility. With respect to the project, Valentina Pellegrini, Vice President of Gruppo Pellegrini, said:

> *As Gruppo Pellegrini serves over 200000 diners every day, we feel a strong responsibility to promote a food culture that considers the needs of individuals who experience challenges with eating.*

The FOOD-AUT project was conducted in partnership with Fondazione Sacra Famiglia Onlus and implemented by the Laboratory of Dietetics and Clinical Nutrition (LDNC) at the University of Pavia (directed by Prof. Hellas Cena). The director of Innovative Services for Autism at Fondazione Sacra Famiglia, Monica Conti, noted:

> *For our organization, which supports over 350 individuals with autism each year, this study offers a valuable opportunity to focus more effectively on dietary habits, which significantly impact overall well-being. By gradually addressing the specific sensory sensitivities of individuals with autism, it is possible to promote more varied, balanced, and healthy diets while enhancing social engagement during meals. The results encourage continued efforts in this direction, including expanding studies to children and adolescents.*

FOOD-AUT aimed to positively impact the health condition of autistic adults, improve food acceptability, and increase dietary diversity. According to our interviewee, the director of Pellegrini's Academy, the purpose of FOOD-AUT was to help autistic people and their caregivers in the delicate and important moment of the meal, as well as provide an additional opportunity to raise awareness of the nutritional needs of this population. On the one hand, it aimed to improve the health status of autistic individuals through the development of menus and recommendations for collective food services to the nutritional and sensory needs of these individuals, as well as the development of nutritional indications aimed at caregivers/parents to support the latter in managing the meals of autistic individuals at home.

The pilot study was conducted at the daycare service of the Sacra Famiglia Onlus Foundation in Cesano Boscone (Italy) between March 2022 and March 2023. Autistic adults (diagnosed according to DSM-5 diagnostic criteria) (DSM-5) [51], which were patients of the Sacra Famiglia Onlus Foundation, were enrolled. Each study participant's parents signed written informed consent.

The FOOD-AUT pilot study consisted of two phases: an observational phase and an interventional phase [52].

The first observational phase of the project involved evaluating the nutritional adequacy and sensory acceptance of the canteen menus for autistic adults and assessing mealtime compliance. Therefore, the nutritional contents (energy, macro- and micronutrients) of the menus provided by Pellegrini S.p.A. at the Sacra Famiglia Onlus Foundation were analyzed via the Italian Food Composition Database (BDA). These values were compared with the nutritional requirements of the individuals on the basis of age, sex, and body measurements. Furthermore, to assess mealtime compliance, meal evaluation forms were used to monitor both the quantity and sensory acceptance of food consumed in the canteen. The quantitative section records the amount eaten via visual and numerical indicators, whereas the qualitative section details the sensory aspects (e.g., color, smell, texture, and temperature). Specific aspects of the foods (e.g., pasta type, fruit peel, vegetable preparation) were also recorded.

Each participant had 12 forms completed over 4 weeks (three nonconsecutive days per week). The menus followed a 4-week cycle, which was repeated three times during the 12 weeks. Forms were filled out by trained staff members, with each operator consistently supervising the same group of 4–5 individuals during mealtimes.

During the second intervention phase, adapted menus tailored to the nutritional and sensory needs of autistic individuals were introduced. The same method of data collection was repeated, with 12 meal evaluation forms collected per participant over another 4 weeks.

Considering the project's results [52], 22 autistic people (29.5 ± 7.1 years old) were enrolled in the study. Most study participants were males (72.7%, n = 16) with a high autism severity level. There was a high prevalence of overweight (54.5%) and obesity (18.2%). A total of 40.9% of the study participants had excessive waist circumference.

The observational phase led to a better understanding of key characteristics associated with greater meal acceptance among autistic individuals, including the following:

- Use of short pasta shapes;
- Low-intensity colors such as white and beige, with bright colors avoided;
- Homogeneity in color within a single dish—preference for ingredients with similar shades rather than contrasting ones;
- Preference for green vegetables over orange vegetables;
- Light-colored or pale fruit (e.g., beige, white), served in pieces;
- Soft or semiliquid textures, with avoidance of foods that are difficult to chew;
- Mild aromas, avoiding pungent or strong smells (e.g., from fish such as mackerel);
- Delicate flavors, characterized by minimal use of spices, garlic, and onion.

With respect to the second interventional phase, the revised canteen menus led to notable improvements in food consumption among autistic adults. Specifically, an increased intake of first and second courses, vegetables, and fruits was observed compared with the "old" canteen menus. This resulted in a better intake of essential nutrients, such as vitamins A, C, D, folate, iron, zinc, calcium, and omega-3 s, which are often lacking in autistic individuals. Water consumption has remained stable, whereas bread intake has decreased due to greater satisfaction with main dishes [52].

Overall, this project demonstrated that tailoring menus to sensory needs can increase food intake, reduce waste, and promote healthier eating habits in autistic adults. Given the promising results, a second multicenter study (AUT MENU) project is currently ongoing to develop national dietary guidelines for this population [53]. The study is conducted by a multidisciplinary team of researchers from the LDNC of the University of Pavia and the Research Centre for Food and Nutrition (CREA) of Rome.[8]

[8] More information regarding this project are available on this website: https://onfoods.it/research-projects/improve-nutritional-and-health-condition-autism-spectrum-disorder-subjects

References

1. Martínez-Lemos RI (2015) Economic impact of corporate wellness programs in Europe: a literature review. J Occup Health 57(3):201–211
2. Eisenberger R, Stinglhamber F (2011) Perceived organizational support: fostering enthusiastic and productive employees. American Psychological Association
3. Ken Research (2024) Global corporate wellness market. Retrieved from https://www.kenresearch.com/global-corporate-wellness-market
4. Gubler T, Larkin I, Pierce L (2018) Doing well by making well: the impact of corporate wellness programs on employee productivity. Manag Sci 64(11):4967–4987
5. Dabos GE, Rousseau DM (2004) Mutuality and reciprocity in the psychological contracts of employees and employers. J Appl Psychol 89(1):52–72
6. Hekman DR, Bigley GA, Steensma HK, Hereford JF (2009) Combined effects of organizational and professional identification on the reciprocity dynamic for professional journal employees. Br Food Acad Manag J 52(3):506–526
7. Guest DE (2017) Human resource management and employee well-being: towards a new analytic framework. Hum Resour Manag J 27(1)
8. Grawitch MJ, Gottschalk M, Munz DC (2006) The path to a healthy workplace: a critical review linking healthy workplace practices, employee well-being, and organizational improvements. Consult Psychol J: Pract Res 58(3):129
9. Dale K, Burrell G (2014) Being occupied: an embodied re-reading of organizational 'wellness'. Organization 21(2):159–177
10. Harvey G (2018) Corporate wellness: what, why not and whither? Empl Relat 41(4)
11. Hull G, Pasquale F (2018) Toward a critical theory of corporate wellness. BioSocieties 13(1):190–212
12. Kavaratzis S, Choi E, Herrewynen M, Bugra T (2023) Investigating the paradoxes of workplace well-being: a systematic literature review. Acad Manag Proc 1:14198
13. Muscaritoli M (2021) The impact of nutrients on mental health and well-being: insights from the literature. Front Nutr 8:656290
14. Koopman C, Pelletier KR, Murray JF, Sharda CE, Berger ML, Turpin RS, Hackleman P, Gibson P, Holmes DM, Bendel T (2002) Stanford presenteeism scale: health status and employee productivity. J Occup Environ Med 44(1):14–20
15. Orfanos P, Naska A, Rodrigues S, Lopes C, Freisling H, Rohrmann S et al (2017) Eating at restaurants, at work or at home. Is there a difference? A study among adults of 11 European countries in the context of the HECTOR* project. Eur J Clin Nutr 71(3):407–419
16. Lachat C, Nago E, Verstraeten R, Roberfroid D, Van Camp J, Kolsteren P (2012) Eating out of home and its association with dietary intake: a systematic review of the evidence. Obes Rev 13(4):329–346
17. Caraher M, Jakšic D, Dolciami F, Stigliani A, Wynne-Jones R (2017) Promoting healthy eating habits in the working population: the FOOD program. MOJ Public Health 6(4):394–399
18. Caraher M, Masanotti G, Bertrand N, Léauté R, Lloyd S, Tavakoly B (2020) Healthy eating promotion for the workplace: the European FOOD (fighting obesity through offer and demand) Programme Promozione di un'alimentazione sana sul posto di lavoro: il programma europeo FOOD (Combattere l'obesità attraverso l'offerta e la domanda). Sistema Salute 64(2):241–253
19. Huang Y, Burgoine T, Bishop TRP, Adams J (2024) Assessing the healthiness of menus of all out-of-home food outlets and its socioeconomic patterns in Great Britain. Health Place 85:103146
20. Mariani M (2024) A paradigm shift in public food service. In: Sustainable food procurement. Routledge, pp 216–229
21. Haddon J (2018) The impact of employees' well-being on performance in the workplace. Strateg HR Rev 17(2):72–75

22. Wanjek C (2005) Food at work: workplace solutions for malnutrition, obesity, and chronic diseases. International Labour Organization, Geneva
23. Mulaney B, Bromley-Dulfano R, McShane E, Stepanek M (2021) Evaluation of workplace wellness programs. Acad Manag Proc 1:12804
24. Saks AM (2022) Caring human resources management and employee engagement. Hum Resour Manag Rev 32(3):100835
25. Allan J, Querstret D, Banas K, de Bruin M (2017) Environmental interventions for altering eating behaviours of employees in the workplace: a systematic review. Obes Rev 18(2):214–226
26. Drewnowski A (2020) Impact of nutrition interventions and dietary nutrient density on productivity in the workplace. Nutr Rev 78(3)
27. Melián-Fleitas L, Franco-Pérez Á, Caballero P, Sanz-Lorente M, Wanden-Berghe C, Sanz-Valero J (2021) Influence of nutrition, food and diet-related interventions in the workplace: a meta-analysis with meta-regression. Nutrients 13(11):3945
28. Schliemann D, Woodside JV (2019) The effectiveness of dietary workplace interventions: a systematic review of systematic reviews. Public Health Nutr 22(5)
29. Page KM, Vella-Brodrick DA (2009) The 'what', 'why' and 'how' of employee well-being: a new model. Soc Indic Res 90(3):441–458
30. Van De Voorde K, Paauwe J, Van Veldhoven M (2012) Employee well-being and the HRM–organizational performance relationship: a review of quantitative studies. Int J Manag Rev 14(4):391–407
31. Adan RA, van der Beek EM, Buitelaar JK, Cryan JF, Hebebrand J, Higgs S et al (2019) Nutritional psychiatry: toward improving mental health by what you eat. Eur Neuropsychopharmacol 29(12):1321–1332
32. Ammar A, Trabelsi K, Hammouda O, Clark CC (2024) Nutrition and wellbeing: how do energy intake, fasting and prudent diets affect mental health. Front Nutr 11:1461415
33. Gross JJ, Thompson RA (2007) Emotion regulation: conceptual foundations. In: Gross JJ (ed) Handbook of emotion regulation. Guilford Press, New York, pp 3–24
34. Evers C, Marijn Stok F, de Ridder DT (2010) Feeding your feelings: emotion regulation strategies and emotional eating. Pers Soc Psychol Bull 36(6):792–804
35. Polivy J, Herman CP (2002) Causes of eating disorders. Annu Rev Psychol 53(1):187–213
36. Thayer RE (2001) Calm energy—how people regulate mood with food and exercise. Oxford University Press, Oxford
37. Telch CF, Agras WS (1996) Do emotional states influence binge eating in the obese? Int J Eat Disord 20(3):271–279
38. Greeno CG, Wing RR (1994) Stress-induced eating. Psychol Bull 115(3):444
39. Polivy J, Herman CP, McFarlane T (1994) Effects of anxiety on eating: does palatability moderate distress-induced overeating in dieters? J Abnorm Psychol 103(3):505
40. Nguyen-Rodriguez ST, Chou C, Unger JB, Spruijt-Metz D (2008) BMI as a moderator of perceived stress and emotional eating in adolescents. Eat Behav 9:238–246
41. Bruch H (1973) Psychiatric aspects of obesity. Psychiatr Ann 3(7):6–9
42. Slochower JA (1983) Excessive eating: the role of emotions and environment, vol 3. Human Sciences Press
43. Scott C, Johnstone AM (2012) Stress and eating behaviour: implications for obesity. Obes Facts 5(2):277–287
44. Payne N, Jones F, Harris PR (2013) Employees' perceptions of the impact of work on health behaviors. J Health Psychol 18(7):887–899
45. Sonnentag S, Pundt A, Venz L (2017) Distal and proximal predictors of snacking at work: a daily survey study. J Appl Psychol 102(2):151
46. Clohessy S, Walasek L, Meyer C (2019) Factors influencing employees' eating behaviours in the office-based workplace: a systematic review. Obes Rev 20(12):1771–1780
47. Patton M (2015) Qualitative research and evaluation methods, 4th edn. Sage Publications, Thousand Oaks

48. Re B, Previtali P (2025) Promoting healthy eating initiatives in organizations: an exploratory study. In: WOA2025 conference proceedings. Associazione Italiana di Organizzazione Aziendale. ISBN 979-12-210-9700-9
49. Kelly RK, Snow S (2019) The importance of corporate wellness programs for psychological health and productivity in the workplace. In: Creating psychologically healthy workplaces. Edward Elgar Publishing, pp 411–430
50. Re B, Zardini A, Sanguineti F, Previtali P (2025) Healthy eating initiatives in the workplace: a configurational approach. Br Food J 127(13):426–444. https://doi.org/10.1108/BFJ-10-2024-1080
51. DSM-5 (n.d.) Diagnostic and statistical manual of mental disorders, Fifth Edition
52. Conti MV, Santero S, Breda C, Basilico S, de Filippo G, Luzzi A et al (2024) Autism spectrum disorder and collective catering service: results of the pilot study FOOD-AUT. Front Nutr 10:1298469
53. Conti MV, Breda C, Basilico S, Ruggeri S, Scalvedi ML, Cena H (2025) Unlocking the link: exploring the association between food selectivity and health outcome in autism spectrum disorder—a systematic review. Nutr Rev nuaf052

Open Access This chapter is licensed under the terms of the Creative Commons Attribution-NonCommercial-NoDerivatives 4.0 International License (http://creativecommons.org/licenses/by-nc-nd/4.0/), which permits any noncommercial use, sharing, distribution and reproduction in any medium or format, as long as you give appropriate credit to the original author(s) and the source, provide a link to the Creative Commons license and indicate if you modified the licensed material. You do not have permission under this license to share adapted material derived from this chapter or parts of it.

The images or other third party material in this chapter are included in the chapter's Creative Commons license, unless indicated otherwise in a credit line to the material. If material is not included in the chapter's Creative Commons license and your intended use is not permitted by statutory regulation or exceeds the permitted use, you will need to obtain permission directly from the copyright holder.

Chapter 5
Concluding Remarks

Beatrice Re

Theoretical Contribution

This book has offered a multidisciplinary and multidimensional perspective on malnutrition, which includes health, economic and organizational perspectives. We aimed to answer the following main macro research questions: (i) How is malnutrition conceptualized, and how does it affect different target groups from a health perspective? (ii) What are the economic impacts of malnutrition? (iii) How do organizations promote worker wellness through healthy eating initiatives? (iv) What benefits do organizations seek to achieve for their employees' well-being and organizational performance through healthy eating initiatives?

By doing so, we provided the following theoretical contributions. First, we offered a comprehensive overview of the global burden of malnutrition, with a specific focus on several vulnerable targets across the lifespan. Second, we offered an extensive explanation of different facets of malnutrition from a health, economic and business organization perspective. Third, we provided insightful case studies of organizations located in the Italian context that implemented solutions to tackle malnutrition; doing so, we offered a contribution to organizational literature on the topic of malnutrition and healthy eating promotion which is to date scant.

B. Re (✉)
Department of Economics and Management, University of Pavia, Pavia, Italy
e-mail: beatrice.re@unipv.it

© The Author(s) 2026

B. Re, P. Previtali (eds.), *Tackling Malnutrition in a Complex World*,
SpringerBriefs in Business, https://doi.org/10.1007/978-3-032-26438-1_5

Managerial Implications

This book provided a multidisciplinary perspective on diet and nutrition. We considered several target groups, including some that have been underinvestigated by the current literature and practitioners alike (such as autistic young adults and organization employees). We suggest that organizations better combine physical and psychological support for their employees to improve both their physical and mental health conditions, which then impacts their functioning in everyday workplace activities.

At the micro level, our findings indicate that investing in healthy eating initiatives within corporate wellness (CW) programs may contribute to improving both employees' well-being and organizational performance. To achieve this, we suggest that organizations may develop effective communication strategies aimed at reducing barriers to the adoption of healthy eating habits in the workplace. Furthermore, we advise firms to establish systematic mechanisms to assess the effectiveness of these initiatives in relation to both individual well-being and organizational outcomes. This evaluation could involve the administration of periodic employee surveys as well as the use of ethnographic research methods. Particular attention should be directed toward worker groups with a higher risk of malnutrition.

Our exploratory study provides a foundation for future qualitative and quantitative research aimed at developing and testing a model on a larger sample of organizations implementing healthy eating initiatives. Such studies would enable broader generalizations across countries and industries. In addition, longitudinal research combining survey data and ethnographic approaches could offer valuable insights into how diet-related interventions influence employee well-being and organizational performance over time.

Concerning the role that organizations may have in addressing DBM and childhood malnutrition, we recommend the implementation of effective initiatives supported by advanced policies (see Chap. 3) within school settings. These initiatives should promote food security, facilitate access to healthy food, and encourage diets that are both nutritionally adequate and environmentally sustainable, while remaining sensitive to the sociocultural context. Moreover, corporate initiatives can support mothers in breastfeeding and in providing appropriate complementary feeding during the first two years of life, thereby ensuring optimal child growth, development, and nutrition [1]. To promote adherence to healthy diets among individuals with low socioeconomic status (SES), organizations may also provide financial support through vouchers for healthy foods—such as fruits and vegetables—redeemable in supermarkets and/or establish partnerships with local farms and shops [2].

With respect to addressing malnutrition in hospitals, healthcare organizations can play a key role across different inpatient (ward) and outpatient settings. For instance, we advise them to screen patients for malnutrition upon hospital admission, and in case of positive screenings, proceed with a formal diagnosis and the development of a tailored treatment plan. In this context, the development and validation of AI-based tools to assist hospital staff in identifying and flagging patients at risk could help reduce the time required for screening. At the same time, we advise hospital managers to design comprehensive and long-term organizational strategies aimed at progressively reducing the structural and operational barriers that contribute to the high prevalence of malnutrition in hospital environments.

To address malnutrition among elderly patients, the recommendations outlined above are also applicable within hospital settings, private outpatient clinics, and nursing homes (the latter showing several risk factors with hospitals that contribute to the development of malnutrition, including institutional routines and therapies, food service practices, and insufficient training of healthcare personnel on malnutrition management).

Suggestions for Future Research

The multidisciplinary research presented in this book opens the way to numerous avenues of research, partially stemming from its inner limitations.

The decision to focus exclusively on high-income countries and to conduct the empirical analysis only in Italy limits the generalizability of the findings and restricts the possibility of exploring the intersectionality between the vulnerability characteristics of the selected target groups and different country-specific factors. Consequently, we call for cross-country studies that can provide a more comprehensive understanding of both aspects.

With respect to the hospital setting, there is an urgent need to increase funding for initiatives addressing malnutrition among hospitalized patients, given its significant negative impact on clinical outcomes. In particular, efforts should focus on overcoming the barriers that currently limit the effectiveness of malnutrition prevention and treatment programs in this context. Furthermore, nutrition programs should be developed and tailored to specific patient conditions in accordance with existing international guidelines. The effectiveness of these interventions should then be evaluated through multicenter studies and, if successful, integrated into routine hospital practice.

Additionally, this book highlights a relatively underexplored population in malnutrition research: autistic individuals. Further studies are needed in this area. Future research should investigate food selectivity and sensory perception more systematically in this population, considering their substantial influence on food preferences and eating behaviors. Studies of neurodiversity in organizational settings could further investigate the relevance of nutritional aspects in designing more neuroinclusive workplaces. To support this effort, the development of an international consensus could help clinicians and practitioners conduct more effective assessments and diagnoses. Moreover, cross-cultural studies would be valuable to examine whether the strategies suggested in the scientific literature to expand the food repertoire of autistic individuals can be generalized across different cultural contexts and appropriately adapted to diverse populations.

From an organizational perspective, the use of a qualitative research design enabled the collection of rich and in-depth insights; however, the findings cannot be statistically generalized. The absence of internal key performance indicators (KPIs) within companies to evaluate the effectiveness of implemented diet-related interventions further complicates the possibility of conducting quantitative analyses. Future research should therefore monitor the implementation of corporate wellness programs over time through longitudinal quantitative studies. Additionally, examining healthy eating initiatives implemented in organizations across different countries could allow cross-country comparisons and facilitate the identification of emerging best practices.

Similar analyses could also be conducted within school settings to explore effective interventions that promote food security and healthy dietary habits. Such studies could identify interventions tailored to the sociocultural context of specific countries to address childhood malnutrition more effectively. Furthermore, future research should investigate the integration of digital technologies, mobile applications, and telemedicine in the treatment of childhood obesity, considering the widespread use of mobile devices among younger populations. In this regard, further studies should examine the role of social media in mental health and its potential relationship with malnutrition, as well as evaluate the effectiveness of integrated interventions targeting both physical and mental health outcomes. In addition, children and adolescents should undergo regular screening by healthcare professionals for the early detection of malnutrition and related comorbidities. Early identification may facilitate timely interventions aimed at preventing the progression of obesity-related complications. Importantly, such interventions should be designed within an integrated care framework and personalized according to individual needs, including mental health support.

Moreover, with regard to populations with low socioeconomic status (SES), future research should prioritize sustainable strategies aimed at mitigating the effects of the climate crisis, expanding nutrition education across all age groups—particularly among mothers—and strengthening social engagement among low-SES populations. Territory-specific initiatives should also be implemented to ensure both the availability and the quality of food. Given the relevance of the socioecological framework of malnutrition, future studies could also examine the impact of workplace food environments, including the availability of healthy options in cafeterias and vending machines, on employees' dietary choices as well as on the eating habits of their children [3].

One last consideration we make concerns the diagnosis of DBM: we argue that basic anthropometric measurements should consistently be assessed and integrated with medical data from clinical records to diagnose undernutrition and/or overnutrition-related malnutrition using validated criteria. In addition, national screening programs for micronutrient deficiencies should be funded, given the lack of updated and comprehensive data on micronutrient deficiencies worldwide.

References

1. Jantzer AM, Anderson J, Kuehl RA (2018) Breastfeeding support in the workplace: the relationships among breastfeeding support, work–life balance, and job satisfaction. J Hum Lact 34(2):379–385
2. Feuerstein-Simon R, Dupuis R, Schumacher R, Cannuscio CC (2020) A randomized trial to encourage healthy eating through workplace delivery of fresh food. Am J Health Promot 34(3):269–276
3. Wickramasinghe K, Chatterjee S, Williams J, Weber MW, Rito AI, Rippin H, Breda J (2021) Childhood overweight and obesity abatement policies in Europe. Obes Rev 22:e13300

Open Access This chapter is licensed under the terms of the Creative Commons Attribution-NonCommercial-NoDerivatives 4.0 International License (http://creativecommons.org/licenses/by-nc-nd/4.0/), which permits any noncommercial use, sharing, distribution and reproduction in any medium or format, as long as you give appropriate credit to the original author(s) and the source, provide a link to the Creative Commons license and indicate if you modified the licensed material. You do not have permission under this license to share adapted material derived from this chapter or parts of it.

The images or other third party material in this chapter are included in the chapter's Creative Commons license, unless indicated otherwise in a credit line to the material. If material is not included in the chapter's Creative Commons license and your intended use is not permitted by statutory regulation or exceeds the permitted use, you will need to obtain permission directly from the copyright holder.

The manufacturer's authorised representative in the EU is Springer Nature Customer Service Centre GmbH, Europaplatz 3, 69115 Heidelberg, Germany. If you have any concerns regarding our products, please contact ProductSafety@springernature.com

Printed and bound by CPI Group (UK) Ltd, Croydon, CR0 4YY
17/09/2026
02218460-0002